Praise for *Generation Restoration*

An urgent and eloquent call to rebuild our bond with nature, and reimagine a thriving future grounded in reciprocity, resilience, and respect.

—Dame Christiana Figueres,
Founding Partner, Global Optimism

Generation Restoration is a timely and inspiring call to action. In a world where every decision—from what we buy to how we live—shapes the future of our planet, this book provides the essential knowledge and tools to reset our relationship with nature. By teaching ecological literacy in clear, practical terms, it inspires individuals and communities to make more conscious choices. I highly recommend it to anyone ready to be part of the solution.

—Ellen Jackowski, TIME 100 Climate Leader

In *Generation Restoration*, Tim Christophersen compellingly illustrates our urgent need—and extraordinary potential—to heal our planet within a single generation. Rich with vivid examples of ecological wisdom and grounded in clear-eyed economic insights, this book is an essential guide to rethinking humanity's relationship with nature, offering both a sobering reality check and a hopeful pathway forward.

—Rhett Ayers Butler, Founder and CEO, Mongabay

Generation Restoration is both a mirror and a map—confronting the hard truths of our current systems while illuminating a bold and optimistic path forward. It challenges our assumptions and shows how regeneration and resilience are core to successful business strategy. A must-read for anyone ready to redefine their relationship with nature for a thriving future.

—Eva Zabey, CEO, Business for Nature

I have had the privilege to get to know Tim Christophersen during our work together on the UN Decade of Restoration Advisory Board these past four years. Tim's long-standing commitment to nature, respecting Indigenous knowledge, and seeking solutions to our climate challenges are visible throughout this book. Anyone interested in a sustainable future should read this; it will make you reflect on your own personal journey and give you new ideas to help our planet adapt to climate change.

—Frank Mars, Member, Board of Directors, Mars, Inc.

Yesterday I toured a solar farm with pollinator-friendly planting between the rows of panels—and as if conjured up from nowhere, insects long thought extirpated from the region have returned to suck the nectar. As this pathbreaking book shows with one inspiring case after another, nature has extraordinary resilience if we give it some room, and sometimes a nudge.

—Bill McKibben, Author of _Here Comes the Sun_

We're in a planetary emergency—a climate crisis that touches every life and every economy. With over half of global GDP at risk from nature loss, the urgency to act has never been greater. _Generation Restoration_ is a powerful, inspiring call to reimagine our relationship with the natural world. Whether you're an environmental advocate, a business leader, or simply someone who cares about the future of our planet, this book will move you to be part of the generation that helps restore our Earth's abundance and resilience.

—Marc Benioff, Chair and CEO, Salesforce

For years, I urged Tim Christophersen: "We need a guide that empowers decision makers and restoration leaders to make courageous, impactful choices." _Generation Restoration_ answers this call with wisdom and humility. Through vivid stories and thoughtful reflection, Tim shows us that hope is not lost—there is still light at the end of the tunnel. What sets this book apart is its deep respect for ancestral Indigenous knowledge. Indigenous peoples have always responded to crises with resilience and proactive action. They planned for a future where their descendants would not suffer the same fate, unlike today's focus on mere adaptation. *Generation Restoration* reminds us that we are part of nature, not above it. It invites us to listen to Indigenous voices, learn from their wisdom, and walk a path of true restoration. For anyone seeking to heal our relationship with Mother Earth, this book is an essential guide—one that inspires both reflection and action.

—Constantino Aucca Chutas, President of Acción Andina and Asociación Ecosistemas Andinos (ECOAN)

As we face a future where clean air, safe water, and a healthy planet can no longer be taken for granted, *Generation Restoration* offers us something deeply needed: hope rooted in action. This is more than just another survey of the issues at hand—it's a love letter to nature, and to humanity. It's a reminder that restoration is not just possible, it's already happening, and WE ARE the generation that must act.

Tim writes with clarity and compassion, drawing from his own rich experiences living in and working amongst local communities—where real change always begins. He reminds us that protecting nature isn't just about preserving beauty or biodiversity, it's about safeguarding our shared future, and recognizing the deep interconnection between people and the planet.

A gift to the next generation of scientists, activists, community leaders, and everyday nature lovers, *Generation Restoration* is equal parts inspiration and roadmap—thoughtful, bold, and deeply personal. With humility and heart, Tim invites us to imagine what's possible when we come together to heal the world we love—and offers a vision for how we get there.

—Jennifer Morris, CEO, The Nature Conservancy

Generation Restoration is a timely and urgent call to reimagine our relationship with nature not as something to exploit, but as a partner in healing our planet. Tim weaves together powerful stories, science, and vision to show that restoration isn't just possible, it's already underway. This book is both a blueprint and a rallying cry for the generation that refuses to settle for collapse. A must-read for anyone who believes in regeneration, justice, and a future we can all thrive in through ecological literacy.

—Kevin J. Patel, Founder, OneUpAction International; Cofounder, Youth Impact Council; American Climate Activist

Tim Christophersen's brilliant narrative in *Generation Restoration* tells us that ecological literacy is the key educational challenge of our times, and ecological regeneration our most important goal because it can unite citizens, businesses and nations in a global transition towards a better and safer future for all life on earth, including our own. Drawing on poignant examples from human and natural history, as well as his rich personal experience, Tim brings both truths to life with memorable examples, crystal clarity, and an admirable sense of purpose. This book is a "must read" for emerging leaders in all walks of life.

—Pavan Sukhdev, Founder/CEO of GIST Impact

Tim's book invites us to return home—to reintegrate with nature. It weaves a compelling scientific and philosophical case for realigning society with the natural world, rekindling ecological literacy, and setting humanity on a new path: the ecological century. It moves on to describe the new economy we need for this project of reunification. An economy that recognizes nature as a source of human well-being, any loss of nature as a liability, and any investment into that most critical infrastructure as an asset. He explains that economics is a mere instrument to index, mint, and share wealth. Assets, currencies, and economies based on nature could make nature internal again to our model of wealth generation—and inspire billions of ecopreneurs to start a century of regeneration.

—Professor Martin R. Stuchtey, Founder The Landbanking Group and SYSTEMIQ

GENERATION

RESTORATION

TIM CHRISTOPHERSEN

FOREWORD BY JANE GOODALL

GENERATION

HOW TO FIX
OUR RELATIONSHIP
CRISIS WITH
MOTHER NATURE

RESTORATION

WILEY

Published by John Wiley & Sons, Inc., Hoboken, New Jersey.
Published simultaneously in Canada.

The manufacturer' s authorized representative according to the EU General Product Safety Regulation is Wiley-VCH GmbH, Boschstr. 12, 69469 Weinheim, Germany, e-mail: Product_Safety@wiley.com.

Trademarks: Wiley and the Wiley logo are trademarks or registered trademarks of John Wiley & Sons, Inc. and/or its affiliates in the United States and other countries and may not be used without written permission. All other trademarks are the property of their respective owners. John Wiley & Sons, Inc. is not associated with any product or vendor mentioned in this book.

Limit of Liability/Disclaimer of Warranty: While the publisher and author have used their best efforts in preparing this book, they make no representations or warranties with respect to the accuracy or completeness of the contents of this book and specifically disclaim any implied warranties of merchantability or fitness for a particular purpose. No warranty may be created or extended by sales representatives or written sales materials. The advice and strategies contained herein may not be suitable for your situation. You should consult with a professional where appropriate. Further, readers should be aware that websites listed in this work may have changed or disappeared between when this work was written and when it is read. Neither the publisher nor authors shall be liable for any loss of profit or any other commercial damages, including but not limited to special, incidental, consequential, or other damages.

For general information on our other products and services or for technical support, please contact our Customer Care Department within the United States at (800) 762-2974, outside the United States at (317) 572-3993 or fax (317) 572-4002.

Wiley also publishes its books in a variety of electronic formats. Some content that appears in print may not be available in electronic formats. For more information about Wiley products, visit our web site at www.wiley.com.

Library of Congress Cataloging-in-Publication Data is Available:

ISBN 9781394328222 (Cloth)
ISBN 9781394328239 (ePub)
ISBN 9781394328246 (ePDF)

Cover Design: Wiley
Cover Image: © Andriy Onufriyenko/Getty Images
Author Photo: © Chloe Jackman

Printed and bound by CPI Group (UK) Ltd, Croydon, CR0 4YY

C9781394328222_260825

Contents

Dr. Jane Goodall's Foreword for Generation Restoration *xiii*

1 Squandering Our Natural Wealth **1**
Restoration in One Generation 2
Nature Gone Bust 4
Shifting Baseline Syndrome 11
Tipping Point 14
Resilience Starts with Us 24

2 A Century of Ecology **29**
Ecology: The Science of Our Only Home 31
A New Century of Ecology 41
Ecological Literacy for Everyone 42

3 Nature Is Us: A Tale of Reciprocity **57**
Nature Is Many, Nature Is One 58
Nature Is Us 61
Counterproductive Conservation 72
A New Hope 77
So, Are We Part of Nature? 83

4 The Value of Nature **85**
"It's the Ecology, Stupid!" 86
A New Economy Emerges 93
The Oldest New Financial Asset Class 99
A Restoration Boom 111
All the Money in the World 115

5   The World Plus 10 Percent   **119**
Take No More Than Half   120
A Movement Is Born: Acción Andina   124
Living in Harmony with Nature: Satoyama   135
The World's Largest Protected Area Network:
Natura 2000   137
The Guardians of Nature   139
Respect for Mother Nature   143
A New Relationship   149

6   A Trillion Trees   **153**
The Power of Trees   154
Resetting Our Relationship with Forests   156
A Rights-Based Approach   162
Trailblazer   165
Learning from the Forest   170
Forests Under Scrutiny   175
Is It Too Late?   178

7   World Restoration Flagships   **183**
Brazil's Atlantic Forest   184
Farther Than the Eye Can See   193
The Arc of Restoration in the Brazilian Amazon   196
Community-Based Natural Farming in India   198
A New Restoration Economy in South Africa   200
Building with Nature in Indonesia   205
Restoring Planet Earth   206

8   Stubborn Optimists   **209**
The Future We Want   210
A Journey Forward to Nature   215
Generation Restoration from Local to Global   227

Epilogue: The Freedom to Choose   229

Taking Action   233

Notes   239

Acknowledgments   265

About the Author   267

Index   269

xi

Contents

Dr. Jane Goodall's Foreword for Generation Restoration

We are going through dark times, politically, socially, and especially environmentally. For thousands of years, early humans, like most animal species, lived in harmony with the natural world, hunter-gatherers taking only what they needed to survive. Gradually, that changed. As our populations grew our demands on the planet's natural resources increased and became increasingly unsustainable. In many instances, *need* became *greed*. Far too many became trapped in a materialistic outlook where success was based on acquiring wealth. There was an unrealistic idea that there could be infinite economic growth on a planet with finite natural resources. Billions of people have become increasingly divorced from the natural world, and instead live in a virtual world defined by technology.

I have spent much of my life studying the amazing animals with whom we share or should share this planet. I have come to understand the complexity of ecosystems, where each animal and plant is interconnected and has a role to play in the complex web of life. The chimpanzees that my team and I have observed and worked to protect since 1960 are amazingly like us. They can live more than 60 years, have distinct personalities, form close bonds between family members, and can use and make tools. They show emotions similar to ours: love, compassion, joy, grief, and so on.

They live in complex communities and are territorial. Like us they can be aggressive and brutal, but like us they can also be loving and altruistic.

There is one major difference that separates us from chimpanzees and other animals—the explosive development of our intellect. For although animals—and not just the Great Apes, elephants, and whales, but rats and pigs, birds, octopuses, and even some insects—are far more intelligent than was once thought, that capacity cannot compare with an intellect that has allowed us to explore outer space and the depths of the oceans and create the internet and AI.

Unfortunately, though we are unquestionably the most *intellectual* creature that has ever lived on Planet Earth, we cannot claim that we are the most *intelligent*—if we were, we would not be destroying our only home. We have lost the wisdom that we see in so many of the Indigenous peoples, who make major decisions only after asking how they will affect future generations. Those who have, for hundreds of years, been stewards of the land.

The good news is that we are beginning to use our intellect to find ways to repair the web of life. As we use our intellect to understand the complexity of the natural world, we are better able to work together to find ways to heal the damage we are inflicting. The path we have been on—one of unsustainable consumption and destruction of nature's resources—has led us to a crisis.

Carbon dioxide and other greenhouse gases have led to a warming planet and changing weather patterns. Species have disappeared at an alarming rate, vast tracts of forests and woodlands have been cleared, wetlands drained, coral reefs bleached, grasslands destroyed. While we cannot fully bring back what is gone, there is much we can do to begin the healing. Nature, when given the chance, has an incredible capacity for regeneration. Forests can be restored, rivers can run clean again, and animals—even those on the brink of extinction—can be given another chance and return to their restored habitats.

Dr. Jane Goodall's Foreword for Generation Restoration

This book, *Generation Restoration*, is a call to action, a roadmap that we can follow as we attempt to heal the harm we have inflicted. It presents a vision of how the world can be for future generations. It is a plea to all—young and old, individuals and nations—to come together to tackle the daunting, but essential task of restoring the Earth's degraded ecosystems on a planetary scale. More than that, it is an invitation for us to reflect on our relationship with Planet Earth, to rekindle a sense of awe and gratitude for the beauty, diversity and complexity of the natural world, for then we will understand the importance of working to protect it and understand that our well-being is inextricably linked to the health of ecosystems-forests, oceans, prairies, wetlands, and all the rest. If we fail, we are doomed. Humans are not exempt from extinction.

It is important to recognize that the movement toward planetary restoration is not so much a scientific or technical challenge, because we know what to do and we have the tools to do it. The challenge is to develop a new mindset in which the protection and restoration of the natural world are central to government policy, business practices, and everyday life. We must reduce unsustainable lifestyles; alleviate poverty; transform industrial farming with its reliance on chemical pesticides, herbicides, and artificial fertilizers; and tackle the problems of pollution and waste, and much more.

It is very important to involve local communities and help them find ways of supporting themselves and their families without destroying their environment, so that they understand that protecting nature is not just for wildlife, but for their own future. For then they become our partners in conservation. I know this is true because of the community-led conservation program of the Jane Goodall Institute in six countries, where we work to project chimpanzees and their forest environment. Of course, in many cultures around the world, the relationship with nature is still strong.

Dr. Jane Goodall's Foreword for Generation Restoration

We have much to learn from indigenous communities as we seek to re-establish a respectful relationship with the world that surrounds us, and on which we depend.

Economic growth, as we have traditionally defined it, can no longer be our guiding star. Instead, we must prioritize the health of the planet and all its inhabitants, and balance this with a way to meet human need and reduce human greed. This will not be easy, but it is essential if we are to create a future where people and nature can thrive in harmony.

Generation Restoration presents a vision of hope, but also a challenge. It is a book for everyone who wants to understand what went wrong in our relationship with nature, and learn how to fix it. Young people, the leaders of tomorrow, are already stepping up to this challenge with passion and determination. They understand that their future is at stake, and they are demanding change. Movements for climate action, conservation, and rewilding are gaining momentum around the globe, driven by a generation that knows we cannot afford to wait—but they cannot do it alone. We must all play our part, recognizing that each of us—no matter how small our actions may seem—can contribute to the restoration of our planet. If enough people, especially those in the corporate world, understand the urgency of the situation, and take action, politicians will support, rather than oppose, the tough decisions that must be made.

Generation Restoration calls on each of us to rethink our relationship with Planet Earth, our only home, and to work together in the greatest restoration project ever undertaken. This is a task for all people of all ages and all nationalities. Tim Christophersen helped to set up the UN Decade on Ecosystem Restoration 2021–2030 while working for the UN Environment Programme. In this book he outlines a collective effort for all of us living in the 21st century, and beyond. In undertaking this journey, we will not only restore the environment of Earth but also restore something deep within ourselves—a sense

Dr. Jane Goodall's Foreword for Generation Restoration

of awe and wonder at the complexity and beauty of the living world around us, and to which we belong.

Over the millennia, Mother Earth has nurtured us, and now she needs our help. Let us move forward with passion, determination, and hope and work together to heal and care for planet Earth. Let us enter into a new era of moral and spiritual evolution.

Dr. Jane Goodall, DBE
Founder—the Jane Goodall Institute
& UN Messenger of Peace

Dr. Jane Goodall's Foreword for Generation Restoration

Chapter 1

Squandering Our Natural Wealth

Any foolish boy can stamp on a beetle, but all the professors in the world cannot make a beetle.

—Arthur Schopenhauer

Thousands of people travel around the world each year and pay vast amounts of money to witness one of nature's most awesome dramas: the great migration on the East African plains, from the Serengeti into the Masai Mara and back. Over 2 million wildebeest and other ungulates and their predators migrate with the seasonal rains, and Earth itself seems to shake with the sheer weight and force of wild animals on the move. Looking on from a safe distance, we suddenly feel small. We marvel at the power of nature. Few people realize that such massive migrations of wild animals once took place all across the world, as recently as a few decades or a few centuries ago.

While living in Kenya for almost 10 years, my family and I witnessed firsthand how nature is still breathtakingly rich. However, we also realized how the former continental-scale migration arena for wildlife such as the African elephant has shrunk to a few haphazardly connected national parks. I remember visiting Amboseli National Park during a prolonged drought and seeing a large herd of elephants, numbering over 100 animals, shuffling around in the bone-dry soil, and kicking up enormous dust clouds. In past centuries, they would have migrated north or west during droughts until they reached a more suitable habitat, returning to the foot of Kilimanjaro only with the next rains. However, in East Africa today, there are few migration corridors left, and many elephant families have lost the experience of migrating long distances and the knowledge of the best routes. The matriarchs of these families once possessed a deep understanding of the migration routes and the wisdom of when to make a move, which they had accumulated over the course of their long, migratory lives. In recent decades, poaching, habitat fragmentation, and human-wildlife conflicts have severely limited their range, not only geographically but also mentally: They have lost the knowledge of how to navigate in landscapes that are now full of fences, roads, and settlements. Their loss of a mental map limits how far they dare to venture.[1] Africa is not the only place where nature is just a shadow of her former diversity and abundance.* Much of our natural heritage has disappeared across the Americas, Europe, Asia, and Australia, and we no longer remember how immensely rich we once were.

Restoration in One Generation

Despite all we have lost, with the right political will, finances, and knowledge, we can regain a diverse and abundant natural world within one generation. This is our generation's moonshot or, rather, our "Earthshot," as Prince William fittingly calls his global initiative to award innovators in this space. Ecosystem restoration is not about a longing for the past. Rather, it is the only way forward that will allow us to enjoy a life in abundance and diversity

---

*Careful readers will note my use of female pronouns, "her" and "she," for "nature" and "Mother Earth." As I discuss in more detail in Chapter 5, using "it" would indicate that nature is a thing or a commodity, which nature most definitely is not.

for generations to come, because our wealth, health, and well-being all depend on nature. We are the first generation that has the global perspective, knowledge, and tools—including sufficient finances—to rebuild nature as our most critical planetary infra-structure. Yes, there are some trade-offs between space for wildlife and space for humans. But there are far more win-win opportuni-ties where more biological diversity creates more abundance for both wildlife and humans. We read about some of those cases in the second half of this book. And ecosystem restoration is as much about people as it is about nature. Countless examples show where the degradation of nature has impoverished people and where ecosystem restoration will trigger a restoration of local com-munities and the economy.

Innovation, imagination, and collaboration are the keys to this collective challenge. Despite the moonshot analogy, there are some key differences between restoring Planet Earth and the Apollo space program. The legendary inventor, architect, and futurist Buckminster Fuller once said that "there is one outstand-ingly important fact regarding Spaceship Earth, and that is that no instruction book came with it."[2] While I appreciate Fuller's quirky humor, Earth is in fact nothing like a spaceship. Earth is more complex, sophisticated, and marvelous than any spaceship we could ever build. Earth is a living system, and we are a core part of that system. The good news is that Earth's systems are not only self-contained but, for the most part, also self-repairing. Earth has immense self-healing powers. To activate them, we need to understand the basic principles of Earth's systems and life cycles. Because, for better or worse, we are the crew of Spaceship Earth, and we need to learn more about this tiny blue and green marble, our only home on which we are hurtling through space. Escaping to Mars is not an option.

Squandering Our Natural Wealth

Before we can move on to build a new relationship with nature, we must take a hard look at the damage we have done and continue to do. Our study of Spaceship Earth begins with an understanding of nature's original wealth and abundance. In the next section, we look at lessons from stories of some ecosystems and species that are disappearing under our watch.

Nature Gone Bust

If you had been a passenger on one of the many ships that took European settlers to the New World in the 17th or 18th century and you were headed to the Caribbean or New York, your ocean passage would have been accompanied frequently by pods of dolphins swimming alongside your vessel and different species of whales spouting at all longitudes—from sperm whales to gray whales, humpbacks, and North Atlantic right whales (named, by the way, for being "just right" for commercial hunting: large, moving slowly, and staying afloat when harpooned).[3] Getting closer to shore in the Caribbean, your sleep would have been constantly interrupted by the thuds of sea turtles crashing into the wooden hulls. The Spanish priest Andrés Bernáldez (1450–1513), a chronicler and contemporary of Christopher Columbus, for instance, wrote in 1494 about Cuba's sea turtles: "The sea was all thick with them, and they were of the very largest, so numerous that it seemed that the ships would run aground on them and were as if bathing in them."[4] So plentiful were these magnificent creatures at that time that ships carried limited provisions of meat because they could easily catch large numbers of sea turtles and store them on board, scooped up from the sea or caught on islands during their journey.[5] Kept alive on board, their meat would stay fresh.

If you were bound for New York and you disembarked at what today is New York Harbor, your eye would have met crystal clear waters that were filtered at least once a day by giant banks of oysters, covering almost 220,000 acres (90,000 ha)—about the size of 180,000 football fields—home to probably more than 300 billion oysters.[6] As one oyster can filter up to 50 gallons (190 l) of water daily, the oysters were continuously filtering the entire coastal ecosystem. They were also feeding the fast-growing local human population. The oyster reefs provided habitat and food to a vast diversity of marine life, including seals, dolphins, whales, crabs, striped bass, and huge shoals of herring. Hundreds of species in the harbor enjoyed the benefits of the vast oyster habitat. The oyster banks almost wholly disappeared after overharvesting, and they have been gone for more than a century now.[7]

In his book *The Big Oyster*, Mark Kurlansky writes:

> Before the 20th century, when people thought of New York, they thought of oysters. This is what New York was to the world—a great oceangoing port where people ate succulent local oysters from their harbor. Visitors looked forward to trying them. New Yorkers ate them constantly. They also sold them by the millions, supplying Chicago, St. Louis, Denver, and San Francisco, but also shipping to England, France, and Germany.[8]

Today, New York has to import its oysters. The city's last commercial oyster bed closed in 1927. Around the world, 85 percent of all oyster reefs have been lost over the last 200 years. Today's oyster populations in the Hudson-Raritan Estuary are less than 0.01 percent of what they once were in New York Harbor.[9] After the last oysters had been eaten and industrial shipping increased, the waters of New York Harbor and the mouths of the East River and Hudson River became murky because of untreated sewage. They are now almost devoid of life. With the disappearance of oyster reefs, the city and coastline have lost one of their most crucial storm and flood defenses. It takes a vivid imagination to see the lost universe of teeming life behind this now-lifeless and polluted body of water. Most people visiting New York today probably think it has always been this way. In reality, the cycle of life has come almost to standstill in the waters around the city.

The Flywheel of Life

Nature basically works like a flywheel, which needs critical mass and speed. Nature becomes more powerful, stable, and resilient the more life and diversity we add. For the past few centuries, since the first Industrial Revolution, most of human efforts have been targeted at slowing down nature's flywheel for the purpose of controlling it. Monocultures, industrial agriculture, and overharvesting of natural resources all go against the fundamental principle in ecology that life and diversity produce more life and diversity. Therefore, the most important thing we can do for ecosystem restoration is to start the process of increasing the diversity and abundance of life. A local initiative is about to do just that for the oysters in New York Harbor. The Billion Oyster Project wants to restore oyster banks around Governors Island and other former sites around the Upper Bay and the river's estuaries. The project releases over 50 million small oysters each season, anchoring them in their former habitat. Millions of oysters have already established themselves in the wild. Despite obstacles such as bad water quality and a lack of funding, the project forges on.

And the momentum is growing, thanks to the enthusiasm of schools, restaurants, ecopreneurs, and investors. The Billion Oyster Project is now part of the curriculum, both in theory and in practice, of more than 100 public school campuses across New York City, mostly middle and high schools, as well as a growing number of schools in northern New Jersey. Once oyster populations become self-sustaining and begin to reproduce on their own in significant numbers, they will be able to form sustainable reefs that support ongoing reproduction and habitat creation. They would start to filter large amounts of water, and their sturdy banks would once again provide storm protection, habitat, and food for fish, birds, and marine mammals. Once we make a strong and intentional start, nature herself will do most of the heavy work. Life could return to the waters around the Big Apple in breathtaking diversity and abundance. Let us now look across the Atlantic at another example of a drastic decline in once-abundant wildlife under our watch.

Meet the European Eel

For centuries, the European eel was the most important commercial fish species in European estuaries. However, following a loss of 95 percent of the entire population over just the past 30 years, the species has recently been categorized as critically endangered in the International Union for Conservation of Nature Red List of Threatened Species.

You might think of eels, if you ever think of them at all, as very slippery (they are!) and perhaps not very good to eat. They are, in fact, delicious, in particular as smoked eel on toast. Eels are medium-sized fish up to 3 feet (1 m) in length and can live up to 20 years in the wild. In my youth, I used to catch eels in a lake near our house. During summer, we would sit around a campfire and wait for eels to take our bait, triggering the little bells we had attached to our fishing lines. To understand what went wrong between us humans and the eel, it is worth a short excursion into the life of *Anguilla anguilla*, which has one of the most fascinating biological cycles in the animal kingdom.

Eels spawn only once in their lifetime, and it was a mystery for centuries where and how they reproduced. In 1922, Danish researcher Johannes Schmidt identified the Sargasso Sea in the western Atlantic as their spawning grounds. The Sargasso Sea is a vast ocean gyre located between Bermuda, the Bahamas, and

7

the southeastern coast of the United States, more than 3,100 miles (5,000 km) from the European coast. When eels sexually mature, their bodies turn a silvery color, and they start their journey from their freshwater and brackish habitats toward the Atlantic Ocean, never to return. Depending on their departure location in rivers, ponds, and lakes across Europe, some eels travel up to 6,200 miles (10,000 km) to get there. They migrate from the freshwater of lakes and rivers toward the coast, into the brackish water of estuaries, and then into the salt water of the ocean. It is assumed that they navigate by Earth's magnetic field. They can breathe through their skin and live on land for extended periods, moving like snakes across fields, preferably during rain or when the fields are still wet from dew. They often live in landlocked ponds and lakes, and to reach the ocean, they sometimes must travel long distances across land to migrate into rivers and streams, which has given rise to many folk tales about their supernatural abilities.

Once they reach the ocean, they start an epic journey westward across the Atlantic, which can take up to one year. During their journey, they no longer eat. Instead, their intestines are transformed into reproductive organs, which are not fully developed before they start their migration. Upon arriving in the Sargasso Sea, they spawn, often at depths of up to 500 feet (2,000 m) and then die. The tiny eel larvae now start an epic journey of their own: They drift with the Gulf Stream eastward, back toward Europe, a trip of up to 300 days. During their Atlantic crossing, they feed on plankton and other marine biomass and grow into small fish, but they are still translucent when they arrive at the southeastern coast of Europe and the Mediterranean, which is why they are called "glass eels" at that stage. From the coast, they start their journey back upstream into even the remotest freshwater bodies of Europe.

The more than 90 percent drop in population size over the past few decades did not happen due to sport fishing, such as my occasional catch of an eel in my youth. Industrial fishing fleets in the Atlantic and Mediterranean catch the translucent glass eels in large quantities, mainly off the coasts of Portugal, Spain, and France, for export to aquaculture worldwide and in particular to Asia, where these eels are considered a prized delicacy. Prices for glass eels soared to 6,800 USD per pound (15,000 USD per kg) in Japan recently. So highly prized is this once-common fish that it has caught the interest of organized crime. The Convention on International Trade in Endangered Species places heavy restrictions on the export of eels, and the annual illegal trade in European eel is estimated to be worth up to 3 billion USD per year.[10] On October 28, 2019, French police arrested two people on their way to Kunming, China, with 300,000 live glass eels in their luggage, in water-filled plastic bags, weighing 200 pounds (91 kg) and worth over 110,000 USD (100,000 euros).

During my lifetime, from my childhood years fishing for eels in our local lake without a care in the world to today, a once-abundant wild animal has turned into a rare and smuggled commodity. Countless other species have met the same fate within the same short time. In the past, fisherfolk or hunters used to just move on to the next species, the next ecosystem, the next exploitation. However, we have reached the end of the road where we can recklessly consume nature to build more financial capital. The blank checks we issue on behalf of nature are starting to bounce. There are few pristine ecosystems left to plunder—the deep sea being one of the few. Yet even the deep sea is no longer safe from our liquidation of its natural capital. A concerted global effort is underway to erode a UN ban on deep-seabed mining, and industrial-scale mineral extraction plans exist across many ocean habitats.[11] The current plan to mine the deep ocean and the trawling of the bottom of the sea for the last remaining fish are in stark contrast to the fact that the ocean was once teeming with life.

Harvest Season

In my home region of northern Germany, much of the year's cultural calendar is based on former animal migrations. The mass migration of herring in their billions along the coast, for example, caused entire villages and towns to shut down for days because everyone was busy catching, drying, and wood-smoking enough fish to last throughout the year. For hundreds of years, the arrival of whales, dolphins, tuna, and sharks, which trailed the herring migrations, triggered a monthlong fishing frenzy among the human populations along the coasts of Northern Europe, North America, and northeastern Asia each spring. Fall and spring also marked the arrival of millions of migratory birds in the Northern Hemisphere. People caught them by the hundreds of thousands in specially constructed duck decoys, and they provided an essential source of protein. For example, in one specially constructed duck decoy in my home region on the island of Fohr in the Wadden Sea, an ecologically essential tidal estuary for migrating birds on the East Atlantic flyway, an estimated 3 million wild ducks were caught and killed between 1730 and 1983.[12] Dozens of such large-scale duck decoy constructions existed along the coast of the Netherlands, Germany, and Denmark.

In late spring each year, when the salmon run began, people would flock to the rivers and catch salmon and other migratory fish, such as the giant Atlantic sturgeon. This once-abundant fish is now critically endangered. Sturgeons can reach 20 feet (6 m) in length and weigh up to 880 pounds (400 kg). The last remaining large sturgeons in my home province of Schleswig-Holstein were caught in the 1950s.[13] Once a common sight and catch in the North Sea and all major European rivers, sturgeons are now so rare that they spawn in only one river in France. The decline of the sturgeon played out over several decades. The collapse of other species took centuries, so we often do not remember how rich our natural heritage once was.

We lack the ability to perceive the loss of nature at an intergenerational pace. Yet, for nature, it happens in a heartbeat. We are losing species too fast for nature to adjust, at an estimated rate 1,000 times faster than the natural rate of species extinctions, but apparently that is still too slow for us to notice. In Earth's geological history, there have been five confirmed mass extinction events, which basically sent evolution back to the drawing board. Some were caused by volcanic eruptions, some by meteorites, and some due to unknown causes. We are now entering the sixth geological mass extinction event, this time caused by us. Even when there is photographic evidence, as in our next story, we don't seem to notice the steady decline of nature until it might be too late.

Shifting Baseline Syndrome

In the early morning of April 14, 1957, three tourists set out on a small chartered fishing boat, the *Gulfstream II,* from the docks of Key West. They were heading for the rich belt of coral reefs off the Florida coast, where they would spend the whole day fishing. For all we know, it was a successful and happy day, judging by the photos taken upon their return, with the day's catch neatly displayed on a special mount at the docks next to the beaming recreational fishermen. The happy hobby fishers and their skipper are posing proudly next to more than 20 large fish of several species hanging from the display board and lying at their feet, including a shark almost 6.5 feet (2 m) long and a goliath grouper larger and presumably heavier than the burly captain.

Over the next 50 years, the same charter company continued to operate fishing tours in the same waters, with the *Gulfstream II* being periodically refurbished. A photo of each trip was taken in the same way, proudly displaying the day's catch. In 2008, Loren McClenachan, a researcher from the Scripps Institution of Oceanography in California, had the opportunity to analyze the entire collection of photos from

1957 to the 1980s, and she took similar photos for comparison in 2007.[14] Her findings are striking. For the 13 groups of most frequently caught trophy reef fish, the average fish size declined by almost 90 percent from an estimated 44 pounds (19.9 kg) to 5 pounds (2.3 kg) over 50 years. Between 1956 and 1960, large groupers and other large predatory fish, including sharks over 6.5 feet (2 m) in length, were commonly caught (see Figure 1.1.). In contrast, by 2007, only small snappers with an average length of 1 foot (34.4 cm) were landed (see Figure 1.2.). Sharks were still caught occasionally, but their average length declined by more than 50 percent. Despite a drastic drop in fish size, variety, and numbers, the company still charged a high price for fishing trips, but customers paid for a much less valuable product.

Figure 1.1 Happy fisherfolk at Key West, Florida, in 1957.
Credit: Monroe County Library

Figure 1.2   Happy fisherfolk in 2007 at the same location, with the average catch size reduced by 90 percent.
Credit: Monroe County Library/Loren McClenachan

What I find most striking about the long-term photo series is that the groups of fishermen and the occasional fisherwoman look equally happy in each photo. They all beam into the camera as if they had just landed the world's best and biggest catch. Imagine if the crew of another charter boat next to them displayed a much larger variety of fish species, at almost 10 times the average size. Our teams would demand their money back. But because there is no immediate comparison to what they have lost, they are blissfully unaware of the steep decline in our natural wealth. They look just as happy in 2007 as in 1957. If this delusion were just a case of a few people's—or even the general public's—complacency with the current state of nature, it might be more humane to keep everyone ignorant about how much we have lost. If you agree, you probably should skip to the next chapter.

A psychological phenomenon called the "shifting baseline syndrome" keeps us buffered from some of the grief or anger we might otherwise feel because of the immense loss of what once was ours and everyone's: seemingly unlimited natural abundance.

However, the shifting baseline syndrome is not merely a mental defense mechanism for coping with the effects of biodiversity loss. It is also a form of self-deception that can lull us into a false sense of security regarding the pace of environmental degradation. It makes us underestimate the looming tipping points at which sudden and drastic changes in ecosystems can occur. It leaves us in the dark about the immense original potential of intact nature and functioning ecosystems to feed, clothe, and shelter humanity as well as to contribute to our recreational and spiritual fulfillment. The shifting baseline syndrome is keeping us poorer, hungrier, and less healthy than we are supposed to be. It is time to reset our original baseline and our future expectations of natural wealth. We should both remember and demand a diverse, abundant world that is rich, fertile, and full of life, because that is how nature is supposed to be and is waiting to be again. Abundance is the true nature of our natural heritage.

Tipping Point

Most of us take the current degraded state of our environment for granted. We redefine what is "natural" with each generation, and our particular state of degradation becomes the new normal. But that is only part of the story. Ecology and ecosystems rarely work in linear ways. Nature moves in leaps and bounds, and Earth's complex life system has many known and probably even more unknown ecological tipping points. To better understand tipping points in nature, imagine a large round boulder resting in a slight depression on a hillside. The boulder represents an ecosystem, such as a forest. When something tries to move and dislodge it, such as a storm or a grizzly bear rubbing against it, the boulder might shift slightly, but the depression it rests in keeps it in place. After minor disturbances, it returns to its resting place. That is called "resilience": bouncing back into a predetermined state of equilibrium. However, when the force moving the boulder is strong enough to push it over the slight edge of its depression on the hillside, the boulder will start to roll downhill with considerable speed and force, until it settles in a new stable location.

The same thing can happen to ecosystems. When their initial resistance to disturbance is overcome and they are pushed over the edge of their natural resilience, they undergo rapid changes, referred to as "tipping points," and their new resting place is farther down the hill of ecological complexity, where they settle into a new equilibrium of birth, growth, and decay. When triggered, such tipping points can suddenly and sometimes irreversibly flip an ecosystem from a diverse and resilient state into a degraded state. The Amazon forest biome, for example, could flip from a moist forest ecosystem into the degraded state of a savannah woodland if it goes beyond approximately 25 percent of deforestation.[15] The warning signs that an ecosystem is close to a tipping point are often overlooked, such as in the case of New York oysters. For much of the 19th century, untreated sewage flowed directly into the waterways, smothering oyster reefs with sediment and toxic runoff. In the early 20th century, New York Harbor was a source of epidemics of typhoid, cholera, and other waterborne diseases due to sewage overflows and industrial waste.[16]

When it comes to the warning signs of our planetary life support systems, we are like the proverbial frog sitting in a pot of heating water. A frog thrown into hot water will immediately jump out again, but a frog that sits in a pool of cold water that is slowly being heated until boiling point will remain there until it dies. (I have not tried this experiment with an actual frog and hope you won't either.) We simply don't realize that we are headed for a point of no return, possibly leading to sudden collapse, because the change is too gradual for

15

our perception within human time frames until it dramatically accelerates. However, we now have clear indications that the intricate web of life is starting to tear at an unprecedented speed. The latest research highlights nine active global-scale ecological tipping points, including ice sheet collapses in Greenland and western Antarctica, the Amazon forest dieback, the permafrost thaw, and the potential collapse of the northern Atlantic Ocean circulation, which provides Europe with its mild climate. A 2025 study estimates a 62 percent average probability of triggering these irreversible tipping points unless we change current policies.[17] We should avoid at all costs letting these systems tip over the edge, which would be irreversible in human time scales. Even before we get there, the great unraveling of our planetary life support systems is already harming even nature's most versatile and resilient species: human beings. And sometimes we cannot return from a tipping point, as these next examples show.

Tipping Points of No Return

Although most of nature's once highly productive ecosystems can be set on a path to recovery—and many of them can even rebound surprisingly quickly, as we see in later chapters—some may be lost forever, or recovery is at least out of reach for many human generations. A study in 2008 predicted that a major population of Atlantic cod near Newfoundland, Canada, would essentially go extinct within 20 years, despite total fishing bans implemented after the collapse of the North Atlantic cod fishery in the early 1990s.[18] For centuries, cod fishing had been the primary economic driver of entire coastal communities. After decades of overfishing in the second half of the 20th century, the cod population suddenly collapsed around 1993, plummeting to less than 1 percent of its original size, which might be too small to recover despite the fishing bans. Even more than 30 years after fishing ceased, the population still shows no signs of recovery. It has gone over a tipping point, and the marine ecosystem has reset itself into a new form of equilibrium. Rolling that giant boulder back uphill seems impossible for the North Atlantic cod, at least within our lifetime.

Another ecological tipping point that is becoming more frequent is the large-scale dying of coral reefs caused by marine heat waves. Spikes in ocean temperature can kill off entire reefs, turning them from lush, colorful havens of biodiversity into monochrome, dead rocks within a few weeks. According to Terry Hughes, a coral reef scientist at James Cook University in Australia, over 54 percent of the world's coral area experienced bleaching-level heat stress in 2023. And the highest ocean heat in four centuries is now even putting the iconic Great Barrier Reef at risk.[19] However, despite clear warning signs, it took the drastic step of UNESCO's World Heritage Commission putting the Great Barrier Reef on the list of World Heritage Sites in Danger to move the Australian government into action. The government is taking additional measures, such as banning gill netting and setting water quality targets. Whether that is too little, too late, or just in time remains to be seen.[20] What is certain is that without drastic emission reductions to slow down ocean warming caused by climate change, the future of the world's coral reefs looks bleak, and their death could happen fast if we let them tip.

Some tipping points unfold over years or decades, which is still lightning fast compared to nature's usual geological time frames. For example, as recently as 200 years ago, an estimated 60 million American bison once roamed the Great Plains of the United States and Canada, feeding and continuously renewing a grasslands ecosystem so rich and productive that it provided everything First Nations peoples needed, feeding hundreds of thousands before the arrival of European settlers. Because the American bison provided much of the food, hides for clothing and shelter, and horns and bones for tools for the traditional lives of Native Americans, military powers of the time recognized that for a clear and decisive victory over resisting tribes,

Squandering Our Natural Wealth

the buffalo had to be destroyed. Deliberate mass killings of buffalo were used as a weapon of war. The vast buffalo population was systematically exterminated until only 541 animals were left by 1889.

In his essay "The Frontier Army and the Destruction of the Buffalo," David Smits writes that the successive generals-in-chief of the US Army, William T. Sherman and Philip H. Sheridan, both recognized that "eliminating the buffalo might be the best way to force Indians to change their nomadic habits."[21] With nature's demise, indigenous peoples could be forced more easily into subjugation and settlements. The land that was thus "freed" by hunters armed with guns and by the military from both buffalo and Native Americans could become agricultural land. However, the large-scale plowing and tilling of the former Great Plains grasslands triggered another ecological and social tipping point a few decades later, resulting in widespread harm to the region's settlers and residents, as we see in the next section.

The Great Dust Bowl

Humanity has learned important lessons from several large-scale ecological tipping points in recent history. One of the best examples of a tipping point caused by human mismanagement, which was later at least partially mitigated and restored by human intervention, is the settling of the American Great Plains Region. The Great Plains covers a portion or the entirety of 10 of the United States: Texas, New Mexico, Oklahoma, Colorado, Kansas, Nebraska, Wyoming, South Dakota, North Dakota, and Montana. These areas were settled by immigrant farmers mostly in the second half of the 19th and the early 20th century. The Homestead Act of 1862 granted 160 acres of land to settlers who would live on and cultivate it for five years. This act, along with the Kinkaid Act of 1904 and others, spurred hundreds of thousands to claim land across the plains. However, the plowing of grasslands, which held the soil in place, greatly disturbed the delicate equilibrium of this vast ecosystem.

In the late 1920s, massive sand and dust storms started to occur over the Great Plains region of the United States, plunging the region into a health and food crisis. Widespread soil erosion due to a combination of severe drought, poor farming practices, and overuse of agricultural soils was causing an ecological disaster that would, over the next two decades, cause one of the largest internal human migrations in US history.[22]

The Great Plains are an ecosystem that has evolved over thousands of years with a balance of grazing, fire, and renewal of grasses. The grasses, such as buffalo grass, are mostly perennials, with deep and complex root systems that could survive prolonged drought, occasional fire, and heavy grazing pressure while at the same time stabilizing the soil. A tiny bit of soil is added each year through dissolving bedrock and through the decomposition and mineralization of organic matter. Forming fertile soil is a process that can take hundreds or even thousands of years, especially in dry conditions, and yet, if the land is not taken care of properly, large amounts of topsoil can be eroded by wind or water in mere hours or days.

In the late 19th and early 20th centuries, the US government encouraged settlers to move west into the Great Plains and plow the grasslands to plant annual crops, such as wheat or corn. This plowing destroyed the deep-rooted, complex underground structure that kept the soil in place. When a severe drought hit the region in the 1930s, the exposed fertile topsoil was swept away by powerful winds, creating massive dust storms. The area hit worst by wind erosion centered around the Oklahoma and Texas panhandles, covering 16 million acres (6.47 million ha) of farmland,[23] an area the size of the US state of West Virginia. The "Black Sunday" storm of April 14, 1935, for example, brought visibility down to near zero and total darkness in the middle of the day across much of the state of Oklahoma.[24]

Squandering Our Natural Wealth

The economic and social damage from this mismanagement of nature remains immense. Based on the assumption that half of fertile soils on an area the size of Iowa were lost in the 1930s and since then only partially recovered, it is estimated that the United States has lost at least half a trillion US dollars in economic output between 1930 and today.[25] This significant, transgenerational loss resulted from the lack of basic ecological literacy among decision makers. Ironically, the prairies could have been sustainably settled by a large number of farmers if they had practiced no-till agriculture and rotational grazing. Instead, the ecological tipping point was quickly followed by a negative social tipping point: As the ecological foundation for society crumbled, lives and livelihoods were severely disrupted, causing widespread human suffering.

Beyond the direct economic damage, the human suffering and property damage caused by the ecological mismanagement of the Great Plains in the 1920s and 1930s were immense. In addition to the erosion of fertile topsoil, the dust storms buried homes, businesses, and roads; killed livestock; and caused widespread respiratory problems, including the so-called dust pneumonia that affected an estimated half a million people during the Dust Bowl.

In his iconic novel *The Grapes of Wrath*, John Steinbeck describes the plight of a family of sharecroppers, the Joads, who leave their home in Oklahoma and migrate to California, where they face severe hardships in the harsh economic climate of the Great Depression. Almost 2.5 million people who migrated during the Dust Bowl from the Great Plains to the American West, mainly to California, may have faced a similar tragic fate.[26] I stress the link among the economy, human suffering, and ecology here because the Dust Bowl is one of the best-documented and most extensive case studies where basic ecological literacy could have saved and improved millions of lives and secured trillions of US dollars in economic value. Numerous similar examples in our past exist from all regions of the world.

Present-day examples include the ongoing destruction of the world's rainforests and destructive overfishing and pollution of the ocean, against all long-term economic reasons. Running our economy without basic ecological literacy is like driving at night without headlights at high speed and hoping not to crash into anything.

On the positive side, the Dust Bowl disaster also ushered in an era of increased understanding of soil conservation and focused government action. Techniques such as contour plowing, planting hedgerows to limit wind erosion, and using native grasses all helped to stabilize soils. The Great Plains Shelterbelt project began in 1934 to stop wind erosion with rows of trees, and, by 1942, it had planted 220 million trees, covering 18,600 square miles (48,000 km²) in a 100-mile-wide zone from the Canadian border to the Brazos River in Texas.[27] All these actions helped to slow the degradation of additional areas. The Dust Bowl offers many valuable lessons for the threat of climate change we face today, including how fast and effective concerted government action can be if we muster the political will. If we learn from the ecological disaster of the 1930s and heed ecological alarm signals early enough, we can repair and maintain the health and functionality of natural carbon and water cycles, particularly in the soil. How we manage soils, given their fundamental role in the global water and carbon cycles, can either stabilize our Earth in its distress or send us even faster into a tailspin. We return to the essential role of soils for humanity in Chapter 8. Let us now dive into the critical role of the ocean for Planet Earth.

Whales to the Rescue

In the 21st century, we may be further removed from an abundant natural world than at any point in human history. Blue whales, for example, the largest animals ever to live on our planet, number only about 25,000 in the world today. Historical blue whale population numbers are hard to quantify. Still, we know that an estimated

360,000 of these majestic animals were killed in the first half of the 20th century alone, almost driving the species to extinction.[28] Blue whale numbers are slowly increasing again since commercial whaling was largely banned in 1986, but it will take many generations for the population to reach more than the current small fraction of about 1 percent of its former population size before the beginning of large-scale commercial whaling in the 17th century.[29]

The disappearance of most blue whales and other large whales from much of the ocean has triggered a ripple effect on the entire marine ecosystem, an effect we have only recently begun to understand. Whales are *ecosystem engineers*: an essential species that shapes entire landscapes and seascapes and on which many other species depend. Whales distribute nutrients and carbon from the depths of the ocean to the surface, and vice versa. They spread essential nutrients and minerals like iron from the relatively small areas of high nutrient upwelling in the open ocean or near seamounts to the vast nutrient-poor ocean deserts that make up the majority of open water. Whales feed in areas where upwelling occurs. When they migrate away from these ocean meadows, they fertilize other areas with feces and urine. In a fantastic feature of over 3.5 billion years of evolution, phytoplankton, the base layer of much of the ocean's food chain, need precisely the mix of nutrients present in whale excrement. This biological stirring of the ocean stimulates and feeds an entire ecosystem of plankton, fish, seabirds, and marine mammals. The whales create islands of life spread across the open ocean.

The loss of whale populations and their migrations across the ocean led to the loss of a critical biological pump of nutrients and carbon, both across the expanse of the ocean and from its depths to the surface. We have lost much of the former richness of many fishing grounds, let alone their vital function of binding carbon dioxide (CO_2) in the ocean depths. Living whales accumulate tons of carbon in their bodies. When a whale dies and sinks to the ocean floor, the dead whale binds many tons of carbon for decades and feeds an entire deep-sea ecosystem. Whales also bring nutrients back up to the surface. Sperm whales, for example, dive up to 6,500 feet (2,000 m) deep, hunting for squid and other cephalopods. And whales spread nutrients when they defecate on their migrations across the ocean. The growth of phytoplankton caused by these nutrients is a highly effective way to draw down CO_2 from the atmosphere. The ocean already absorbs almost one-third of all annual CO_2 emissions through the photosynthesis of phytoplankton, drawing carbon down into the ocean depths when the plankton dies and sinks. Whales were once the primary catalysts for this marine cycle of life by distributing the nutrients for a vast carpet of phytoplankton across the ocean as the foundation for much of marine life. The absurdity of our common human mindset of scarcity versus the reality of nature's approach of abundance is demonstrated by the fact that, to this day, in some areas of the world, whales and dolphins are hunted because they are seen as competitors to human fisherfolk. In fact, they are one of the main reasons that there are fish in the first place.

In a 2023 study, Heidi C. Pearson from the University of Alaska and other scientists estimated that the direct and indirect carbon sequestration, fixation, and storage contribution of the world's great whale population each year is more than the equivalent of 90 million tons of CO_2, or roughly the same as the annual carbon sequestration of Germany's entire forest area. However, the same study estimates that historic whale populations, which were significantly larger, helped to absorb over 30 times more CO_2, a stunning 2.5 gigatons of CO_2e each year.[30] (A gigaton is 1 billion tons, an enormous volume of CO_2.) In other words, about 5 percent of all our current annual greenhouse gas emissions could be eliminated if we allowed whale populations to reach their historic numbers again. Only a few nations still hunt whales, and most whale populations show signs of

Squandering Our Natural Wealth

recovery. Unfortunately, whale species reproduce relatively slowly, and it will be many decades, if not centuries, before we reach a point where whales once again can give life to the entire open ocean and draw down carbon at a significant scale. The whale pump shows the immense power of nature to regulate and run Earth's carbon cycle—and the enormous risk we are taking by allowing our ecological life support systems and wildlife populations to plummet to their current levels. Our global economy seems to be on a self-destructive autopilot, aiming to extract as much monetary value from the living world as possible, until nothing is left. That autopilot has led us to disregard that we are losing altitude fast and are headed into dangerous territory. We can still take back the controls of Spaceship Earth before it is too late. Instead of escaping to Mars or plundering the deep sea, the last large, untouched ecosystem on Earth, let us rewrite the next chapter of our human drama and give it a happy ending, starting with ecosystem restoration.

Resilience Starts with Us

Even if you don't care about European eels or other recent examples of dwindling wildlife populations caused by overfishing, overhunting, or habitat destruction, you might care about the indirect impacts that our disregard for nature and ecology has on our lifestyle. The supply chains of everyday commodities, such as coffee, cocoa, and timber, are starting to disintegrate. After three years of bad harvests in West Africa due to drought induced by climate change, cocoa prices reached almost 10,000 USD per ton in 2024, five times the price of 2020. Other agricultural commodities are coming under similar pressure across the world. To continue to thrive as a civilization, we have to replenish our natural "bank account" and start to live off the interest that accrues from nature rather than burning through humanity's only true source of wealth and abundance.

Fortunately, with a recent global move beyond gross domestic product and toward measuring the true value of nature as part of inclusive national wealth and a growing movement to invest in nature conservation and restoration, this shift is starting to happen, as we see in Chapter 4. Nature-friendly and regenerative growing techniques, such as agroforestry or shade-grown cocoa and coffee, exist for many of our agricultural products. Regenerative agriculture holds the opportunity of higher nutrition, more farmer income, and greater food production on restored soil. We return to the topic of agriculture and our globalized, industrialized agri-food system throughout this book, because that system is the main driver of nature loss. At the same time, agriculture has the potential to become the greatest driver of nature restoration. Due to our current agri-food system and other drivers of nature loss, we live at a fraction of the natural wealth and health we and life on Earth are entitled to.

A 10 Percent World

Across all ecosystems and all biomes, we have lost most of the world's natural abundance in just the past 500 years. Today, only about 4 percent of the world's weight of mammals are wild animals; the rest are livestock and humans.[31] Since 1492, the world has been in an accelerating race to use up, or "liquidate," nature for short-term economic gain. We have burned, eaten, and chopped our way through much of our natural wealth: clean water, clean air, forests, wildlife, and fish. So much, in fact, that the share of nature per person on the planet has dropped by over 40 percent just between 1990 and today.[32] An estimated 66 percent of ocean ecosystems are now damaged, degraded, or modified, and one-third of all commercial marine fish populations are fished unsustainably.[33] Natural abundance, wildlife populations, and biodiversity today are only a shadow of the world's former natural wealth. In his wonderful essay "A 10 Percent World,"

J. B. Mackinnon explores the emerging field of historical ecology.[34] He argues that we live in a "10 percent world" when comparing today's natural abundance with Earth's recent geological past.

This statistic is not meant to be precise. From the perspective of the many species that have gone extinct since humans emerged, we live in a zero percent world. The Intergovernmental Science-Policy Platform on Biodiversity and Ecosystem Services estimates that up to 1 million species may be threatened with extinction.[35] Some wildlife populations today are well below 10 percent of their natural range, and some are even below 1 percent. And a few are starting to make a cautious comeback, such as beavers, bison, and wolves in Europe, or the bald eagle in the United States, which had a phenomenal comeback through better protection after near extinction in the 1970s.

It is hard to determine a precise percentage of loss across all types of ecosystems and all species. Coral reefs are dying back at record rates and might completely disappear with global warming reaching 2°C or more, the upper limit for global warming that was set in the Paris Agreement on Climate Change, somewhat arbitrarily, as a "safe limit" for human civilization to operate on Planet Earth. Yet more than 50 percent of the world's original forest cover still exists from before the Agricultural Revolution, thankfully, and forest area is on the increase in Europe, China, and other regions, even though forests are still being lost at an alarming rate in most of Africa and South America.[36]

Providing a precise global statistic for the status of all of nature is not the point here. As J. B. McKinnon points out, "Imagination is not a game best played between columns of data. What is the taste, smell, and feel of a wilder world?" We currently live on a fraction of nature and our own potential, and I use the term "10 percent world" as a metaphor for our self-inflicted nature poverty. Living in a 10 percent world severely curtails human creativity, health, and well-being. It clearly limits our economic and social development.

We can move from a 10 percent world—a world where only a skeleton of our natural riches remains, and we are picking at the bones—to living once again in a 100 percent world. It is the only way for us to maintain a life in abundance in the long term. Fortunately, there is some cause for optimism.

Staging Nature's Comeback

One piece of good news is that nature conservation and restoration are highly effective when done right. Recently, there have been signs of some species recovering, particularly mammals. A study in 2023 by the Zoological Society of London, BirdLife International, and the European Bird Census Council revealed that of 50 European species studied (25 mammals, 24 birds, and 1 reptile), almost all showed an increase in population size and range. The most stunning recovery was observed in European beaver populations. Once hunted almost to extinction, the beaver population has increased by 16,000 percent (160 times) over the past 30 years in the 95 locations that were studied.[37] Similarly, Eurasian wolves, European bison, and red deer are all making a comeback. At the global level, a study in *Science*, the world's most respected scientific journal, showed that conservation efforts are working in two out of three cases when the researchers reviewed evidence from 665 trials from 186 studies conducted between 1890 and 2019 across various countries, oceans, and species.[38] Although that is good news for some mammal species, we still see downward trends in insect populations, with an up to 75 percent drop in the last five decades across Europe, and half of all the world's bird species are now in decline.[39,40] We still have a long way to go. But we can—and we will—rebuild our natural world. Restoring nature at a planetary scale is an idea whose time has come.

It is clear that the world of the future will look very different from today's world. Our future world could either become, by degrees, worse: warmer, less diverse, and more violent both in weather extremes and in human interactions. Or the future world can become orders of magnitude better than today's world: more resilient, more connected, more diverse, healthier, and wealthier. This positive change will be made possible by an intentional choice to embrace the power of nature and combine it with human ingenuity, technology, and political will to create a world of abundance. Which is the future we choose? How can we get from a 10 percent world to a 100 percent world? Once we have realized that we are undermining humanity's most important relationship, how do we make peace with nature? Before we attempt to answer those questions, let us dive deeper into the story of one of the most underrated scientific disciplines of our time: ecology.

Chapter 2

A Century of Ecology

We must cultivate a deep sense of gratitude for the Earth,
recognizing all that it provides and all that it has taught us.
　　　　　　　　　　　　　　　　　　—Robin Wall Kimmerer

In March 2024, I attended a conference in Livingstone, Zambia, at the iconic Victoria Falls. More than 500 ecosystem restoration experts from around the world had convened in a beautiful new hotel on the banks of the Zambezi River. Our Nature-Based Solutions conference was mostly going well. However, the power went off several times each day in the hotel. That is not unusual in parts of Africa and other regions. Still, it was uncommon for Zambia, which has a stable and ample supply of hydropower that provides more than 80 percent of the country's electricity. But in the spring of 2024, the country was experiencing a record drought. It had not rained properly for several months, and President Hakainde Hichilema had just declared a national emergency. The usually reliable Kariba Dam, which is fed by the Zambezi River, came close to shutting down for the first time in its 65-year history. To make matters worse, it was expected that more than 70 percent of the country's annual food crops would fail due to a lack of rain.

When faced with a potential breakdown of energy supply and food security, any government will shift into crisis mode. Suddenly, questions of ecology and nature restoration rose to the top of the president's agenda, as the country's future depended on them.

The Zambezi River's only remaining water supply originated from a few intact mountain forest watersheds, where the deep roots of trees store rainwater in the soil, releasing it as a steady, clean flow over time. And the only farmers in Zambia who still expected a substantial harvest were those who had shifted from monoculture field crops to multistory and diverse agricultural systems, incorporating crops and trees, a method known as agroforestry. Right after the conference, President Hichilema invited the Global Evergreening Alliance, the conference organizer, to develop a border-to-border plan for ecosystem restoration in Zambia, as he rightly saw it as the only viable way to secure the country's future.[1]

I am sharing this story with you because it is what every country could be facing in the coming decades unless we address our relationship crisis with nature. Nature can easily cause a breakdown of any human society: She can shut us off from water, energy, and food overnight. The impacts are coming closer for all of us, even for those of us fortunate enough to live somewhere still somewhat protected from nature's imminent collapse and from the increasingly extreme weather events caused by a changing climate. This crisis has deep root causes, and one of them is our lack of understanding of our true relationship with nature. Our knowledge and application of ecology, the science of our common home, will determine whether we prosper or suffer in the years to come.

Examining our relationship with nature is the central focus of this book. Why does our reliance on nature not figure more prominently in the minds of politicians, investors, and other decision makers? Nature underpins all our activities, both as individuals and as our societies at large. Despite that dependency, we have allowed ourselves to slide into a relationship crisis with nature. For many people across all walks of life, nature is just an afterthought—if they ever think about nature at all. As we have with a stable global climate, we have taken functioning natural systems and processes for granted because nature has been there for us for as long as humanity can remember. But both the abundance and health of nature and the stability of the climate are now crumbling all around us.

To fix our relationship crisis with Mother Nature, we first need a basic understanding of our common home, which is the ancient Greek meaning of the word "ecology." We need to know our options now that humanity is starting to crash into the so-called Planetary Boundaries—a set of scientific metrics that delineate our safe operating space on Earth against nine essential life support parameters, including a stable climate, freshwater quality and quantity, and biodiversity.[2] We need to renew our understanding of nature's ability to sustain all life and our own essential role as planetary ecosystem engineers. So, in this chapter, we look at how Planet Earth works.

Ecology: The Science of Our Only Home

Ecology studies the relationships between all living beings, including people, and their environment. It has long been a neglected field of science. Recently, however, much more research into the functioning of our Spaceship Earth is being undertaken, and the latest discoveries and research tools, such as artificial intelligence, make it clear that Earth is not a spaceship at all. It is unlike any machine or mechanism we know. Earth is a living, incredibly complex system of ever-evolving relationships among genes, individual life-forms, species, and their habitats, and even between different energetic states of matter. I suggest we examine a few basic ecological parameters and bust a few myths about nature before we turn our attention to a new relationship between us and nature and to restoration of nature at large scale.

Suzanne Simard and other authors, who discovered a complex underground network of communication, synergies, and exchanges of nutrients across forest ecosystems sometimes referred to as the "wood wide web," triggered a recent shift in ecology.[3] Simard's research, which numerous other studies have since confirmed, shows that collaboration between species is far more common in nature than competition. Darwin's "survival of the fittest" view of competition between and within species as the primary driver of evolution has become much more nuanced. The discovery of complex interspecies relationships in forests has led to the understanding that plants, like animals and every form of life, have a form of consciousness, albeit one that is very different from our own. This understanding is revolutionizing our view of agriculture and the immense importance of biodiversity within healthy soils as the main driver of plant growth. All of life exists within ecological communities. No being is an island. To truly understand nature, we should view it as a complex, interconnected entity, or a living system, if you prefer a more scientific term, rather than just a collection of individual species. And that entire collective called "nature" includes us.

Earth itself might be one giant, self-regulating organism of sorts, according to James Lovelock's Gaia theory.[4] The Gaia hypothesis proposes that life on Earth has actively shaped the planet's environment, creating optimal conditions for its own survival. In other words, our climate as well as the global cycles of life—of water, carbon, and nutrients—are largely a result of the interaction among living organisms, Earth's crust, and the atmosphere. Ecologists question the Gaia hypothesis, but it provides a useful mental image of nature as an indivisible whole, with us intricately connected, rather than as a collection of bits and pieces of genes, species, and ecosystems. We do know that we urgently need to stabilize the climate and Earth's major life support systems, such as the ocean, and to do that, we need to repair ecological functions and restore nature on a large scale. We are currently far from planetary-scale nature restoration, but it is possible, and in fact, it is starting to happen, as we see in Chapters 6 and 7.

Perhaps it takes a national crisis like the one in Zambia in 2024, or similar ones unfolding across the globe, from inundated cities to parched fields to outbreaks of new diseases, to remind us that we and our technology are not entirely in charge of Gaia's self-regulating system. However, although we might not be in charge, we are probably the most important biological factor in Earth's self-regulation. It is high time to remember our place and responsibility within Mother Nature's broader family and play our proper role in the cosmic story of our planetary community. We have the power to change our actions and become a positive force for the planet instead of causing harm. To do this, let's look at some of the most important events in Earth's natural history. Right now, we are leaving the Holocene era, which began approximately 11,700 years ago at the end of the last glacial period. We are entering the Anthropocene—a new era of uncertain duration when humans are the main influence on the planet's future—and we are not off to a great start.[5]

Ecological Fever Pitch

Let us use the Gaia hypothesis as a useful mental image for a while and assume that Earth is a self-regulating system of biodiversity, geology, and the atmosphere, much like the human body is an interconnected, self-regulating system of cells, bones, and organs.

For a basic understanding of planetary ecology, let us further imagine for a moment that the planet is a human body.[6] The average body temperature of Planet Earth in the 20th century has been around 57°F (13.9°C), quite a bit cooler than the temperature of a human body. Soil and ocean make up the planet's skin, and, just as with the human body, it is our largest organ and one of the most important ones. Like the human body, Earth's biosphere—the thin layer between the upper atmosphere and the deep ocean that can sustain life—consists mainly of water. For life to exist on Planet Earth, there must be the right mix of moisture and temperature, just as in the human body. Now let's look at the impacts of climate change on our planetary body.

When I speak with people about climate change, the 2.7°F (1.5°C) target set by the Paris Agreement and the 3.6°F (2°C) global warming maximum established as the safe upper limit, I often hear that these sound like minor differences, measured in tenths of a degree. What is the big fuss? Let us return to our body of Planet Earth and our average temperature. Even 2.7°F (1.5°C) of global warming would equal a more than a 10 percent increase in Earth's average temperature. If this happened to the human body, our temperature would increase from 98.6°F (37°C) to 105.2°F (40.7°C)—enough to cause a severe fever and perhaps require hospitalization.

However, increasing Earth's temperature by 20 percent (the equivalent of our current global warming trajectory of 4.8°F/2.7°C) would be the same as the human body temperature rising from 98.6°F to 111.9°F (44.4°C), which would cause almost certain death. And let's keep in mind that the 4.8°F (2.7°C) scenario is what the Intergovernmental Panel on Climate Change calls its "intermediate emissions scenario," meaning it is not even the worst case.[7] The current best-case scenario is 2.5°F (1.4°C) warming by the end of this century, which is approximately the same level of warming we are experiencing now—and which I would argue is already outside of safe limits. Unlike the human body, life on Planet Earth is a vast, highly diverse, and resilient system that has survived countless fever bouts and ice ages. It is unlikely that life on Earth would disappear completely even at a much higher temperature increase. But what would undoubtedly disappear is the stability and predictability of the system on which all human civilization has been based and built over the past 10,000 years. Our entire infrastructure—everything from roads and airport runways, railroad tracks across permafrost soil, and cities built near the sea— is built for a very narrow and stable climatic band. And our civilization is built on the assumption that nature will always provide an ample supply of goods and services, such as clean drinking water and fertile soils that provide us with food.

Even small changes in these basic life parameters will throw human civilization into turmoil. More important, changes in any living systems, in both human bodies and entire ecosystems, do not always follow linear, predictable pathways. Nature works in leaps and bounds, with sudden, drastic tipping points. The differences between 2.7°F (1.5°C) and 4.8°F (2.7°C) of warming might sound small and linear—a little bit worse with every tenth of a degree.

In reality, these warming scenarios represent the difference between some coral reefs possibly dying due to marine heatwaves and almost all of them certainly dying. Or the difference between the Amazon rainforest perhaps flipping to a new ecological state of a savannah woodland to it certainly becoming a frequently burning savannah, and turning from an ecosystem that absorbs and stores carbon to one that starts to emit carbon. Or the difference between just some of the West Antarctic ice shelf melting and all of it going into irreversible meltdown, raising sea levels globally by up to 17 feet (5 m) over the coming centuries.[8] Or the difference between catastrophic fires hitting California every 100 years or every 100 weeks. All these changes could be triggered by global temperature differences measured in tenths of a degree of global warming. A small increase on paper equals giant differences for humankind and our fellow Earth inhabitants.

Uncharted Territory

Limited natural climate fluctuations have occurred over the past 10,000 years, and they have always had major impacts on human history.[9] However, the current scale and speed of temperature change are unprecedented. The first decades of the 21st century have been warmer than any sustained period since about 125,000 years ago. Even more concerning is the speed of the current temperature rise compared with anything that Planet Earth has seen. The fastest natural climatic transitions of our geological past caused by the heat-trapping greenhouse gases carbon dioxide (CO_2) and methane were much slower than our current human-caused greenhouse gas emissions and the resulting temperature rise. The last time CO_2 levels were as high as today was approximately 3 million years ago, during the Pliocene geological period, with CO_2 concentrations in the atmosphere of around 400 parts per million (lower than the levels recorded at the time of writing this book). Earth's average temperature during the Pliocene was about 5.4°F (3°C) warmer than it is today, and sea levels were as much as 60 feet (20 m) higher.[10] It is misleading that most of our climate models only run until the year 2100, when in fact we might be triggering irreversible effects that could impact humanity for thousands of years. Our global response to the climate crisis has not yet risen to the level of the danger we face or the burden we could place on countless future generations.

The good news is that we are not mere bystanders watching the global carbon and water cycles, our two main currencies of life on Earth, lose their balance and vitality as they transition from sources of life to sources of destruction. As Earth's most powerful ecosystem engineers, we can play a significant role in helping the planet once again self-regulate its climate. Restoring ecosystems on a planetary scale, particularly tropical rainforests and the world's agricultural systems, can help cool the planet significantly, as we discuss later in this chapter and in Chapter 6. The world needs to focus more on reducing carbon emissions and phasing out fossil fuels as the main causes of climate change. At the same time, restoring Earth's carbon and water cycles through ecosystem recovery can be a fast and efficient concurrent solution to combat climate change. And restoring nature at scale will have benefits for all of humanity well beyond climate change mitigation.

The current script for the human drama on Planet Earth is dire and does not point toward a happy ending. But we can change our role on the planet and our relationship with nature. To give us some inspiration for what our new role could look like, let us take a quick look across all of human history and pick out some of the best and worst examples of humanity's relationship with nature.

A Century of Ecology

As German poet Goethe said, "They who cannot be far-sighted / Nor three thousand years assay / Inexperienced stay benighted / Let them live from day to day."[11] Human civilizations have encountered many ecological crises before, and we always moved on—sometimes by design, though more often by disaster. We can learn from history and understand how to repair our relationship with nature, as well as how to restore ecosystems on a large scale, before it is too late.

It is, of course, a tall order to have several thousand years of human history at our fingertips or even in the backs of our minds. Fortunately, technology can help. Recently we have been able to look farther and more clearly back in time than ever before, thanks to the development of new archaeological techniques, the discovery of previously unknown historical scripts and other sources, and the assistance of artificial intelligence. There have been examples of ecosystem restoration turning around the fates of entire civilizations.

In his seminal book, *A Forest Journey*, John Perlin traces the intertwined fate of humanity and forests over 5,000 years from ancient Mesopotamia to the 21st century.[12] The historic trail of our dependence on nature includes a period in history that could be called a "century of ecology" in ancient Greece, roughly from 400 BCE to 300 BCE. Several centuries of overuse of forest resources had resulted in the complete loss of ancient Greece's naval superiority by 400 BCE. By the time the country's archrival, Sparta, finally defeated the Greeks in the Peloponnesian War (431–404 BCE), Athens had exhausted its domestic supply of wood for warships, whose complex design required thousands of trees each. Athens had used up its strategic reserve of 100 warships in its fruitless attempt to win the war. By about 400 BCE, the city-state experienced such acute wood shortages that even the unauthorized collection of fallen branches for firewood was severely penalized. There was no wood fuel to heat homes, or often not even to cook meals, other than for the ruling class. Ancient Greece was facing a relationship crisis with nature.

Restoration to the Rescue

Against the backdrop of an acute ecological crisis, philosophers Plato and Aristotle triggered a mind shift in the public's perception of nature.[13] Plato's *Republic* describes a near-perfect imaginary city-state, Kallipolis, ruled by philosopher-kings.[14] In the utopian Kallipolis, Plato frequently uses nature as a metaphor to explain human behavior, governance, and justice. For instance, the metaphor of the soul's harmony parallels the balance found in a healthy ecosystem. The *Republic* describes a transition from the "healthy city" to the "fevered city," which is dominated by overconsumption, complexities, and conflicts. The demand for more resources led to the need for expansion, which ultimately gave rise to war. This critique can be seen as a warning against the overexploitation of nature and the consequences of greed and unsustainable growth. Although Plato does not explicitly discuss ecological degradation, his concern for the balance and sufficiency in the "healthy city" of Kallipolis underscores the importance of living within natural limits.

Around 400 BCE in Athens, the scarcity of natural resources affected every aspect of daily life. In a society and economy that depended on wood for everything from energy to raw materials, the rampant deforestation made a wiser use of natural resources inevitable. Against this backdrop of ecological crisis, Aristotle developed the concept of *eudaimonia*, or human well-being, which can be interpreted as a call to balance human individual needs with the needs of society and human flourishing with the health of nature.[15]

Aristotle's view that virtue lies in moderation and balance suggests a philosophical foundation for sustainable practices in natural resource management. As only about a third of all Aristotle's written work has survived, he may have weighed in on the public debate at the time regarding how ecology should inform the economy (or, in other words, how the knowledge of *oikos*, our common home, should influence the management of nature and society). As the most prominent student of Plato, Aristotle's opinion would have had a strong influence. Both Plato and Aristotle wrote and spoke about the need for the wise management of nature. In the *Laws*, for example, Plato discusses the importance of managing land and resources sustainably.

Rebound

When the *Laws* was published, groundwater levels around Athens were dropping, and many wells had run dry. The rich but fragile topsoil had been washed away in many parts of Greece by flash floods from the deforested mountains. The once-densely forested hills around Athens contained "nothing but food for bees" by 400 BCE. Plato's utopian scenario may have helped the people of Athens and all of Greece understand the importance of ecology, how much natural riches they had already lost, and how to get them back with a concerted public and private effort.

It is unclear how much influence specific public policies promulgated by Plato and Aristotle had, or whether the century of ecology from around 400 to 300 BCE was simply a consequence of the level of natural resource extraction, making it necessary to invest in restoring ecosystems at a national scale. What is clear is that new public poli-cies were established to restore forest ecosystems, and innovations by private citizens and companies in energy-efficient construction brought firewood consumption to significantly lower levels, allowing for the widespread recovery of forest resources. By about 300 BCE, forests around Athens and much of Greece had recovered enough to allow the city-state to take once again a prominent place among the leading Mediterranean powers of the era, thanks to ecosystem restoration and the principle of reciprocity with nature.

Why has the world forgotten the most basic lesson of ecological literacy—that our well-being as a human society depends entirely on nature? We do not have Socrates, Plato, and Aristotle around today to help us focus our minds on the need for a reciprocal relationship with nature. But surely we can still access their writing and wisdom. In addition, we have a much more powerful arsenal of knowledge and tools at our disposal, including artificial intelligence, which could help us identify and design the best possible environmental policies from across all of history and all countries. We can design and guide satellites to monitor the enforcement of nature conservation and help reward the implementation of conservation and restoration actions with new financial tools, including blockchain and mobile pay sys-tems that could allow for direct incentives to citizens for restoration efforts. None of this was possible to the ancient Greeks, and they still succeeded. So can we.

Another key ingredient for success today is the growing number of innovative *ecopreneurs*: young (or young-at-heart) entrepreneurs who make it their business and their life's mission to solve human-ity's most pressing environmental problems while making a decent living. We need to rediscover some basic ecological principles to allow modern solutions to flourish. Indeed, ancient Greece, as well as other past civilizations, can teach us how we can avoid the most dangerous pitfalls and how to achieve national and planetary-scale ecosystem restoration faster. And although speed is of the utmost importance, what we need even more urgently is a clear long-term perspective and a positive vision for humanity.

A New Century of Ecology

What if a positive vision for humanity drew upon the best from 5,000 years of our past, such as the century of ecology 2,400 years ago, and projected our collective dreams, hopes, and aspirations for many generations into the future? Not just a few years, as is the case with most governments, or just three months, as is the case with most businesses filing quarterly earnings reports. What if humanity could collectively learn from our past to shape a new role and responsibility on Earth? Coming out of the current UN Decade on Ecosystem Restoration (2021–2030), we could embark on a new century of ecology, where we properly understand and repair Earth's ecosystems and rebuild our relationship with nature based on interdependence and reciprocity.

The United Nations has laid the groundwork for this in the UN Decade strategy, which states:

> Although the UN Decade on Ecosystem Restoration ends in 2030, it aims to create a platform for societies globally to put their relationships with nature on a new trajectory for centuries to come. It is envisaged that this trajectory will include: nature being respected across society; ecosystem restoration taking place over hundreds of millions of hectares and generating millions of new livelihoods; human rights, with a focus on gender equity, youth, local communities, Indigenous peoples and future generations, being central to restoration initiatives; global supply chains and consumption patterns being shifted to protect, sustainably manage and restore nature; long-term scientific research

being used to guide restoration initiatives; and the value of nature being a central pillar of national systems that assess economic well-being.[16]

I address some of the requirements and benefits of the Decade on Ecosystem Restoration and a century of ecology in the chapters to come, in particular the role of Indigenous and local communities; the value of nature in economic decision making; and the opportunity of deeply personal, inward change for each of us that has ever been the only way to change the world. The first step on that inward journey is particularly exciting: gaining new knowledge. We can learn to read and heed Earth's emergency signals and reconnect with nature across all levels of society and enjoy the benefits of a world of abundance. Put simply, we all need to become ecologically literate.

Ecological Literacy for Everyone

Despite Buckminster Fuller's quip about Spaceship Earth discussed in Chapter 1, reading the distress signals of our home planet is not rocket science. Climate science, for example, can seem highly complex, where supercomputers are used to crunch models that involve petabytes of data (a petabyte corresponds to approximately 500 billion pages of standard printed text). However, every farmer and local community member now knows that the climate is changing and that the rapid change causes problems. It does not require a PhD in climate science or ecology to understand the crisis we are in, and we should not limit the discussion about our collective response to expert circles. We should grow the group of people who are observing and listening to the critical signals from nature. We can all see and feel the impacts if we pay attention, and we can all be part of the solution. By learning basic ecological literacy, we can empower everyone on Planet Earth to read, hear, and understand nature's language.

Nature is broadcasting on all frequencies, all the time; all we need to do is stop and listen.

The lessons we can distill from listening to and observing nature are the same that have kept many Indigenous and local communities alive and thriving for millennia, as Chapter 5 explains. By acknowledging and actively striving to be part of nature again, we open ourselves up to being nature's voice for any decisions that affect our common home. Nature needs us as allies right now. Despite the overwhelming evidence that we depend on nature, we currently treat Earth, including the complex food web beneath our feet, at best like a machine that requires a few chemicals and energy inputs to produce a particular commodity as an output. The perceptions of Earth as a spaceship and nature as a commodity have reached their limits. Earth is a complex living system, just like we are. Just as we need to know a few basic things about how the human body works, we also need to know a few basic things about how our planet, or Mother Nature, works. After all, we are all part of that larger system.

Ecological literacy is not as difficult as it sounds. Just like the ability to read and write, it takes an earnest attempt to learn, the right teachers and tools, and some perseverance. As we are all part of nature, we are already born with an innate understanding of natural systems and the interaction between species. According to E. O. Wilson's biophilia hypothesis, humans possess an inherent, biologically based affinity for nature and other living organisms.[17] This connection is rooted in our evolutionary history, as humans evolved in close contact with natural environments. We just need to open our minds and senses to rediscover the wonderful lessons nature has in store. Becoming ecologically literate can start with answering four simple questions for ourselves: Where does my food come from? Where does my water come from? What is nature doing for my neighborhood? And which fellow species live around me?

Where Does My Food Come From?

As with many good relationships, our relationship with nature starts with food. What we eat and drink daily matters not only in terms of being ecologically aware. It is also the most critical lever for changing our relationship with nature and halting and reversing the degradation of ecosystems worldwide, as the UN Decade on Ecosystem Restoration calls for. So, let's start with food and water. Our globalized agri-food system accounts for approximately one-quarter of all greenhouse gas emissions, and it is responsible for more than half of all land-based biodiversity loss. Agriculture is a good place to start, because once we are ecologically literate, we can turn agriculture from a carbon source to a carbon sink and from a destroyer to a steward of nature and a champion of ecosystem restoration. And change starts with where you live and what you choose to eat.

Do you know where your food was grown? Who are your farmers, where are they based, and how do they grow your food? Fortunately, you don't have to travel around the world to find out. Many consumer goods labels can tell you how and where your food was grown. And your exploration of the origins of your food should start close to home. If you have nearby farms or farmers markets and can purchase fresh, preferably organic produce directly from the producer, it would certainly be highly appreciated not only by the farmers but by you and your family. The basic principle of eating regional and seasonal food will be good for your health and your wallet. The world is too large and diverse to give any advice that can apply to everyone when it comes to food. However, there are a few basic rules, such as eating food that is as natural as possible. I don't advocate for not eating any meat, because livestock, and especially cows, are essential for regenerative farming and soil health. With the right management, cows and other large herbivores can draw carbon into the soil, as we see in Chapter 8. "It is not the cow, it's the how," as my friend Peter Byck, producer of the fabulous *Roots So Deep* science documentary, would say.

However, cows and other animals we raise for food are converting energy and using water that could otherwise be directly consumed by humans, so, in general, people should decrease their meat consumption and increase the quality of the meat produced through better animal welfare. The choice my family has made is to eat meat in moderate quantities and only when it is from animals that have lived as natural a life as possible, preferably locally produced. As we have our own sheep, ducks, and chickens, some of it is indeed very local. Another basic rule of thumb is to adjust our cooking calendar to the seasonal and regional cycles of available fruits and vegetables. That is, of course, easier if you live in the tropics—we enjoyed year-round fresh fruit when we lived in Kenya—but even in temperate climates, a surprising diversity of local produce is available year-round. And if it is not, ask for regional produce in your local grocery store, or try to discover local farm stores or farmers' markets. In case you think this sounds more expensive than buying highly processed foods or fast food, you might be surprised how fairly priced locally produced and directly marketed food can be, especially when you consider the benefits of good food to your health and to the health of nature. We examine the true cost of food most closely in Chapter 4.

Once you dive into the simple question "Where does my food come from?," I hope you will find the answer educational and entertaining, and ultimately beneficial for your health, finances, and overall well-being. Strawberries in January in the Northern Hemisphere not only have a dubious taste and a considerable CO_2 footprint. They are also quite expensive. I am not advocating for specific actions or restrictions. Food preferences are highly individual, and building a deeper and more conscious relationship with nature will naturally guide you toward a diet that is better for you and for the planet.

Gaining a different perspective and understanding of our relationship with nature likely will trigger numerous beneficial changes, including more awareness about the many values of good food. Responsible farming as a profession should be held in much higher esteem by all of us. We literally depend on farmers for our lives, and they in turn depend on the health of the soil and on functioning ecosystems.

My wife and I realized a lifelong dream when we purchased a piece of land a few years ago to start a small farming operation, aiming to become as self-sufficient as possible and gradually expand our production for sale. Although our farming adventure is highly gratifying, it is also demanding and has high initial costs, so I would only recommend it to those who genuinely love gardening or farming. My respect for family farmers has grown by leaps and bounds. Done in an integrative way, farming is one of the most knowledge-intensive and complex of human endeavors—though it is also one of the most rewarding. No matter where you start, once you feel more like you are part of nature where you live and work, you naturally might consume less (or, at least, higher-quality) meat, eat more regional and seasonal products, buy less processed foods in favor of more fresh food, and become more conscious of the circularity of all the key processes in nature that our food system depends on. This change will not feel like an effort or a sacrifice. It will be a win-win for nature and you, and it will feel right. We return to the nature of food in Chapter 8. Let us now turn our sight on water, arguably the most essential ingredient for life on Earth.

Where Does My Water Come From?

The next question is where your drinking water comes from. More than one-third of all major cities in the world receive their drinking water from forest watersheds,[18] and more than half of all humans now live in cities. For example, cities such as New York, Bogotá,

Johannesburg, Tokyo, and Vienna obtain most of their drinking water from protected forest areas. Yet probably few of us would be able to point out on a map which forests store and filter our drinking water. A notable exception is Kenya, where citizens have been educated for many years about the national importance of the country's "water towers," the main mountain watersheds that are the source of most of the country's drinking water, including Mount Kenya, the Aberdares, and Mount Elgon. The freshwater flowing from these ecosystems is essential in more ways than one: More than 20 percent of Kenya's electricity comes from hydropower from dams along the water towers,[19] and the rivers originating in the mountain ecosystems provide irrigation for much of the country's agriculture.

How can you learn where your water comes from? Contact your local water utility to find out. In our family's case, it comes from a borehole located near our house that is only about 39 feet (12 m) deep and lies close to a neighbor's conventionally managed agricultural field. Every time we see the farmer spray pesticides on the crops of wheat or potatoes—and nonorganic potatoes need a lot of poison to grow, apparently—we are concerned about the quality of our drinking water. We test our water regularly, and, fortunately, no pesticides have been detected in our well so far. Still, the nitrate levels are already above EU norms and significantly higher than US safety limits, which are more stringent. Nitrates are widely used in fertilizers to improve crop growth and are also a standard component of explosives. Dissolved in water, they have no color, smell, or taste. Nitrates run off fields and farms into rivers and streams, infiltrating drinking water and potentially impacting health, particularly the health of newborns. If necessary, we will need to dig a well closer to home. However, we are also trying to convince our neighbors to adopt organic agriculture or, even better, regenerative agriculture, which can achieve high yields without industrially produced nitrates and other potentially harmful chemicals.

How Does Nature Help My Neighborhood?

Another illuminating aspect of ecological literacy is discovering the direct dependencies, risks, and benefits from nature. For example, if you live on a hillside or your city or village is close to the sea or a river, which ecosystems can protect you from severe rain or flooding? If you live in an urban area, do you have any trees nearby (or, if you are lucky, in your yard), and do you know how much they can lower average summer temperatures? Trees save lives during urban heat waves. It is good to remember that mature trees evaporate entire bathtubs full of cool water from deep beneath the soil on a hot day, assisting with cooling entire neighborhoods. And they filter and absorb air pollution particles, including heavy metals, thus keeping them out of your lungs.[20]

Unfortunately, trees are not evenly distributed in most cities. American Forests, one of the leading think tanks and implementers of large-scale forest restoration in the United States, systematically mapped tree distribution for the first time recently for the United States and the United Kingdom.[21] Its "Tree Equity Index" across 150,000 different American neighborhoods and 486 metropolitan areas shows how income and other social indicators are linked to the availability of green spaces. The results are sobering. Neighborhoods with the highest income levels have more than double the tree cover per person compared to less affluent communities.[22] Less affluent areas with significantly fewer trees are missing out on essential health benefits, such as cleaner air. To achieve tree equity, the United Kingdom would need to roughly double its average urban tree cover to 30 percent. To achieve tree equity across the United States, 522 million trees need to be planted in urban areas.[23] Fortunately, American Forests and other organizations lobbied successfully for federal funding to increase forest cover in US cities, and, in 2023, a historic 1.5 billion USD was made available to start a massive urban greening campaign.[24] Trees and forests are also refreshingly nonpartisan. Citizens from across the political spectrum usually can agree on the importance of urban trees, and public funding is still flowing for this important investment.

Which Species Live Around Me?

Once we know our direct dependencies on nature in terms of our food, our home, and our health, the next element in ecological literacy is to know something about the fellow species near us. If you live in a city, you might think there is no nature near you, but even big cities have more plants and animals than we assume. For this step in ecological literacy, we can use the latest technology, which is easy and fun. Numerous apps can help us get acquainted with our animal and plant neighbors. Merlin, for example, is a birdsong recognition app that utilizes smartphone microphones to detect and identify most songbirds worldwide. Get up early one morning, go to a nearby park or green space, and hit record. Unless you do this in the middle of winter or the middle of a complete concrete or agricultural desert, you should be able to meet and greet a few species of birds.

With plants, it is even easier. For example, iPhones have an in-built plant recognition function, and there are apps, such as iNaturalist, that can tell you (with fast-improving accuracy, thanks to artificial intelligence) what species you are looking at. Other apps, such as LeafSnap, can help you become a budding botanist. This online ecological literacy training can become more sophisticated as you progress. For example, I subscribe to an app called Mushroom ID, which helps me on my mushroom-hunting trips to our local forest—although the app clearly states that you should not eat any mushrooms based only on its suggested identification. Soon AI will further enhance our ability to recognize nature, and we might all have AI agents with us, like PhDs in our pockets, to identify any plant, animal, or fungus. But as we learned earlier, nature is more than just a collection of individual species. It is a complex system of interdependent processes. Let's take the next step in our ecological literacy journey: understanding the primary life cycles of nature.

Rain Makes Forests, and Forests Make Rain

We humans tend to think in linear ways: "I do this, then that happens, and I might gain what I want." Rarely do we consider all the consequences of our actions and their knock-on effects over an extended period of time. In contrast, almost any natural process is circular, self-correcting, long term, and cyclical, at least when ecosystems are in balance. For example, we all know it takes some rain to ensure that a forest can grow. But did you know that forests make their own rain and probably even their own wind to transport it?

Imagine tropical rainforests as giant natural air conditioners, which work through a process called "evapotranspiration." Evapotranspiration is the combination of water evaporating from the surface of leaves and soil and water vapor transpiration from plants. This water recycling system of forests, particularly rainforests, is key to regulating the entire planet's water cycle. For example, each tree in the Amazon rainforest evaporates up to 264 gallons of water (1,000 l) per day. There are an estimated 390 billion trees in the Amazon basin, representing 16,000 different tree species,[25] and they evaporate water so effectively each day that continental-scale amounts of water are sent through the air, in so-called flying rivers that bring rain to much of South America.[26]

However, the Amazon is at the brink of a dangerous ecological tipping point, where the entire ecosystem could tip from being a moist tropical forest biome to a semidry forest savannah. Scientists have estimated that this shift into a new ecological state could happen rapidly if about 25 percent of the current forest area is cut.[27]

Deforestation now stands at close to 20 percent of the entire Amazon biome. Ecological literacy at the national level, not just at the individual level, is necessary. For example, understanding the ecology of rainforests and their critical role for continental-scale weather and precipitation patterns can trigger fiscal policies that transfer funding into forest conservation to protect rainfall for agriculture.

Fortunately, the Brazilian government recently reduced the deforestation rate by almost 50 percent; in 2024 the rate was one of its lowest levels in the past decades.[28] Other Amazon countries, such as Peru, Bolivia, and Colombia, have also seen a recent reduction in deforestation rates. And although conservation of the Amazon remains a priority, it is no longer enough. We also need forest restoration on a massive scale, which is now being organized, as we see in Chapters 6 and 7. It is a tight race against time and against an ecological tipping point that may occur at any moment.

Forests as Continental Air Cooling Fans

It has long been known that forests produce rain through evapotranspiration by recycling large amounts of rainfall back into the atmosphere. However, the *biotic pump model* suggests that forests not only generate rain but even influence wind at a continental scale. The biotic pump concept is a fascinating ecological theory that further highlights the importance of forests in regulating the world's micro- and macroclimates, beyond serving as giant carbon "vacuum cleaners," or carbon sinks.

According to the biotic pump theory, forests create low-pressure zones when water vapor from trees' evapotranspiration condenses into clouds, and this low pressure draws in humid air from the oceans, even over great distances.

When large sections of forest are destroyed, the biotic pump weakens, leading to less rainfall, more drought, and eventually a shift toward desertification in places that once received ample rainfall. The concept is still contested but is supported by other recent discoveries, such as the flying rivers of massive water vapor streams that originate over the Amazon rainforest. If proven correct and if forests indeed are creating the wind that carries the moisture to bring rain far into the interior of continents, this scientific concept would have far-reaching consequences. China, for example, receives about 80 percent of its rain from the west, all the way across Europe and Siberia, due to the flying rivers that carry moisture over the vast Siberian forests. Would China be willing to help fund the conservation and restoration of the Russian forests that it depends on? Should Brazilian and Argentine farmers, who have grown rich from soybean farming, be taxed to restore sufficient forests in the Brazilian Amazon and the Atlantic Forest along the east coast of Brazil to keep the biotic pump going to water their crops?

Save the Forest, Save the Rain

Since the 19th century, an area roughly the size of China or the United States has been deforested. If we were to restore a significant portion of these forests worldwide, global temperatures could lower and erratic weather patterns could be stabilized. A recent study suggests that regenerating about 1 million square miles (approximately 2.5 million square kilometers) of tropical rainforest, an area roughly the size of Argentina would be sufficient to prevent the planet from warming further, giving humanity time to complete the energy transition away from fossil fuels, a process that has begun and is accelerating.[29]

The math aligns with research from the University of Zurich at the lab of ecologist and charismatic innovator Thomas Crowther. In 2024, lead author Lidong Mo in Crowther's team, together with over 200 other scientists, published an updated version of a 2019 paper in *Science* that confirms that restoration of forests at a planetary scale can indeed slow and even reverse global warming if sufficient emission reductions accompany it. The authors estimate the total global carbon sequestration potential of forests at about 226 billion tons (226 gigatons), about four times as much as our current global annual emissions.[30] They also provide detailed suggestions where the conservation and the massive forest restoration could occur without threatening food security by replacing too much agricultural land.

Restoring vast areas of forests the size of Argentina or more might sound daunting. And, of course, we also must stop further deforestation as we invest in turning the tide and restoring forests on a massive scale. Two key aspects are crucial to consider when contemplating planetary-scale forest restoration. First, we would not necessarily need to turn agricultural land back into forest land. About one-third of our existing forest land is degraded and could be restored to a higher level of ecological functionality. And even within existing agricultural land, there is space for billions of new trees, such as food, fruit, or shade trees, in agroforestry systems. We learn hear more about global forest restoration in Chapter 6.

Second, this investment could secure humanity's collective future and would yield numerous benefits beyond stabilizing the climate, which puts the costs into perspective. It's not as expensive as it may sound. If we take a conservative estimate of global average restoration costs of 2,400 USD per acre (6,000 USD per ha),[31] then restoration of 965,000 square miles (2.5 million km^2) would cost 1.5 trillion USD, about 2 percent of the world's current annual gross domestic product, stretched over a decade or more. That is less than the world already spends on the impacts of climate-related extreme weather events.[32] To me, it sounds like a bargain for saving our collective future. Restoring ecosystems is not just about cooling the planet and slowing climate change. It's about reviving the biosphere, enhancing biodiversity, and creating a healthier, more resilient world. And as an emerging and labor-intensive industry, ecosystem restoration at scale could create millions of new jobs around the world.

As we see in Chapter 4, the money is available; we just have to redirect it from being spent on the wrong things, such as state subsidies for oil and gas. We are approaching a point where we will have no choice but to take bold measures and redirect massive amounts of public and private finance to stabilize the global climate and mitigate nature loss. In the process, we will achieve a stronger, more resilient economy; secure food, water, and energy in perpetuity; and create millions of green local jobs.

Tropical Forests Forever

The idea of restoring forest ecosystems and landscapes at a planetary level would have sounded like an unrealizable dream even a few years ago. However, in the fall of 2023, at the UN Climate Summit in Dubai, the president of Brazil, Luiz Inácio Lula da Silva, proposed the Tropical Forests Forever Facility, which aims to finance the conservation and restoration of tropical rainforests at levels previously unthinkable. His idea of incentivizing tropical countries to conserve and restore forests is gaining momentum—and, it is hoped, just in time. The flying rivers of the Amazon already show signs of drying up. A record drought plagued the entire Amazon in 2024, with people walking across rivers that usually are filled with water year-round. The proposed 125 billion USD Tropical Forests Forever Facility would pay tropical forest countries for maintaining and increasing their forest cover, as a service to all of humanity.

By restoring forests, wetlands, rivers, and agricultural systems, we can enhance water cycles, increase the planet's vegetation cover, and draw more carbon into plants and soils. The solution to cooling the planet may not be some futuristic technology but a return to the natural processes that have always sustained the carbon cycle on Earth to allow all life to flourish.

In fact, about one-third of all our climate mitigation needs can come from nature, together with numerous other benefits, such as clean water, biodiversity, and a sustainable supply of nature's products such as food, timber, and nontimber forest products. In a report commissioned by our team at the UN Environment Programme in 2021, titled *Nature-Based Solutions for Climate Mitigation*,[33] we reviewed all major studies published to date on nature's potential in helping stabilize the climate. We estimated that if all necessary measures for the protection and restoration of global ecosystems are taken, nature could draw down an additional 11.7 gigatons of CO_2 per year from the atmosphere, about twice the annual emissions of the United States and about half of the current global "emissions gap" needed to stay within the 2.7°F (1.5°C) target.[34] About 62 percent of this potential for nature-based climate solutions would come from forests, about 24 percent from solutions in grasslands and croplands, and 10 percent from peatlands. The remaining 4 percent would come from the protection, management, and restoration of coastal and marine ecosystems.[35]

Nature as a Multisolver

Nature already has most of the solutions we need—from slowing climate change to providing our entire society and economy with everything we need in abundance and in perpetuity. We just need to allow nature's ecosystems to recover and play their stabilizing and provisioning role at a planetary scale. This recovery will not happen overnight or in a few years. The UN Decade on Ecosystem Restoration has just gotten us started. The decade was approved by the UN General Assembly and all UN Member States in March 2019. Over 400 partner organizations and more than 100 countries worldwide are now working on ecosystem restoration and a new relationship with nature. The UN Decade can lead us into a century of ecology, restoring this planet to levels of natural wealth that we have all but forgotten, enabling long-term human well-being beyond our current ecological crisis.

Earth thinks in eons rather than in years. We can make a strong start toward a restored Planet Earth by fixing our relationship crisis with Mother Nature in our generation. Doing so will require continuous adjustments and improvements over time, as any healthy long-term relationship does. But what exactly is our current relationship with nature, and how can it be improved? How do we fit into Planet Earth's amazing web of life? In the next chapter, we explore our true human nature.

Chapter 3

Nature Is Us: A Tale of Reciprocity

*We have forgotten that we ourselves are dust of the earth; our
very bodies are made up of her elements, we breathe her air
and we receive life and refreshment from her waters.*
—Pope Francis, *Laudato Si'* (2015)

The central question of this book is this: How can we fix our
relationship crisis with nature? As we see in this chapter, we
have neglected or forgotten our natural identity or even have actively
denied it. The realization that humanity cannot exist outside of or
separate from nature is both necessary and overdue. We are an indivisible part of nature. Every degree of separation from nature we
hold in our minds translates into real-life damage, over time, to ourselves and to our planetary ecosystem.

As we saw in Chapter 2, nature is a living system, not a machine
or technology we fully control or even yet fully understand. Within
this system, everything is connected with everything else all the time,
just as all parts of our bodies are connected. And we are an integral
part of this system. The illusion that humanity exists outside of nature
is increasingly dangerous. Fortunately, the predominant worldview
that humanity, with all our technology, creativity, and culture, is separate from nature is starting to change. Let's explore the emerging
realization of our true relationship with nature.

Nature Is Many, Nature Is One

The realization that there is no absolute separation between us and nature is the most challenging part of ecological literacy, because this concept is so different from our usual cultural context and from most of human technology, where every aspect of our lives, every project we carry out, and every social structure we are part of has clearly defined boundaries. The intersubjective realities we create for our societies to function, such as the concept of nation-states or other social groups, usually have very clearly defined boundaries, at least in our imagination. In the case of nation-states, the borders we draw on a map are real to us, even though they exist only as social constructs. It is useful to be aware that the boundaries we draw around our social groups, such as family, community, or nation, also exist primarily in our minds.

To understand nature, we need to think holistically rather than about small blocks or entities within nature. We tend to use our mental scalpel of compartmentalization when viewing nature, separating it into specific entities. For example, we categorize different parts of landscapes or seascapes into "habitats" or "ecosystems," which sound like clearly defined geographic areas or functional units. However, if you ask ecologists how to determine the exact boundaries of an ecosystem, to a degree where it is no longer connected to or dependent on anything else, they would tell you that it cannot be done. I heard my favorite example of the fluid border between ecosystems when I once visited a village in the rainforests on the island of Borneo. The many trees in the gardens and along the roads created a seamless transition from the village into the forest. When I asked my hosts where their town ended and where the forest began, they said the rule was that as long as you could hear a rooster cry, you were in the village, and when you could not hear it anymore, you were in the forest.

Let me further illustrate the concept of the ever-evolving and ever-present interconnectedness between all parts of nature using the biotic pump concept, which we introduced in Chapter 3. Tropical rainforests thrive in the Amazon because it rains there approximately 6.5 to 10 feet (2–3 m) per year. At the same time, the main reason it rains a lot in the Amazon is that almost 400 billion trees live there. There is no real separation between the rain and the forest. Together, they form the rainforest, and countless habitats and species have emerged over the millions of years during which this system has evolved toward diversity, complexity, and abundance. This web of life would not exist in the same way without any of its parts, and those parts, in turn, would not exist without the system as a whole. The same can be said for all life and all species, including us.

The Jenga Game

A fair amount of redundancy is built into nature, meaning that usually many similar species can coexist in similar habitats under similar ecological conditions. That principle of redundancy creates stability and is nature's secret to success for surviving and thriving during times of rapid change. It creates an enormous pool of individuals, populations, and genes from which to adapt and evolve in response to the ever-changing flow of life through the millennia and geological eras. However, the concept of resilience based on redundancy has limits. It is somewhat similar to the Jenga game, where players remove individual pieces from a tower of wooden blocks until the entire structure becomes too unstable and collapses. The game inspired artist Benjamin von Wong to create his breathtakingly beautiful statue "Biodiversity Jenga," the main attraction at the UN Biodiversity Summit in October 2024 in Cali, Colombia (see Figure 3.1). The 20-foot-tall statue is constructed

Figure 3.1 The "Biodiversity Jenga" statue at the UN Biodiversity Summit in 2024: Humanity is perched atop the intricately balanced tower of nature. With every building block removed, our position becomes more unstable. Will we have the wisdom to replace the blocks in time?

Credit: Benjamin von Wong, 2024

from differently colored and illuminated boxes that represent the various species and ecosystems on Earth. As humanity removes box after box, we realize only when the tower has become precariously unstable that the world's children are sitting at its very top. Will we let them fall? Or will we put back the necessary biodiversity building blocks to stabilize it? The statue draws attention to the fact that we are more than just dependent on nature; we *are* nature. Why does this profound truth not play a more critical role in our everyday language and everyday lives?

Nature Is Us

Humanity in the 21st century seems to live by a kind of philosophical autopilot, programmed about 300 years ago by leading Western thinkers of the so-called "Enlightenment" Era and further based on ancient interpretations of religious wisdom.* The thinkers and religious leaders of that time viewed humans as the only beings on Earth with a direct link to God; everything else was just an intricate machine or commodity placed here to support us. During my university years and for some time thereafter, I was also an unquestioning believer in the credo that humans were the absolute and undisputed rulers over all other species and Earth itself. Perhaps I was just too busy to think much about it, because as soon as you start to ask "Are we part of nature?" it quickly becomes clear how odd and outdated the notion is that *Homo sapiens* should be separate from all the rest of creation.

It is time to examine the "Enlightenment" Era teachings about our role in nature and consider whether they still work for us, now that we have eroded and degraded nature to such a degree that it is crumbling all around us. Whether we are part of nature seems a simple question, but it requires a nuanced and thoughtful answer. As we see in this section, how we define our relationship with nature is more of a moral choice on our part rather than an obvious scientific conclusion. We are just starting to understand the many ways in which life is connected with other life, how our cells communicate within our bodies, and how life communicates within and across

---

*In my view, the term "Enlightenment" is a misnomer for this era. Although it brought sweeping changes by challenging traditional authority, some of its philosophies—such as a disdain for nature—are actually counterproductive to true enlightenment, understood as the pursuit of greater understanding and wisdom.

species. Even though we lack complete ecological knowledge about all aspects of life, the question of whether we are part of nature can be answered on several levels. Part of the answer has a clear foundation in our current understanding of natural sciences. Each of the levels discussed next brings us deeper into philosophy, mythology, and, finally, faith.

Human Taxonomy

Let's start with the zoological facts. We are bipedal primates with a large brain. What sets us apart from other primates is our highly developed capacity for language and our talent to make complex tools, thanks to our opposable thumbs and large brains. In biological taxonomy, we are neatly categorized within the class Mammalia; the suborder Primates; the infraorder Simiiformes (higher primates); the family Hominidae; the subfamily Homininae; the tribe Homini; the subtribe Hominia; and the genus *Homo*. Our species is called *Homo sapiens*, the "wise ape." Our closest living relatives are the other great apes. Based on our biological taxonomy, the case seems clear: We are just another species.

And yet no one with any pride in human intelligence or achievements would concede that we are simply another animal. The level of our abilities and intelligence seems so far beyond that of other animals that it can be hard to imagine we have much in common with them. Yes, we share more than 95 percent of our DNA with most other primates and approximately 80 percent even with mice. But we can split the atom, send someone to the Moon, and construct artificial intelligence. Yes, we are related to animals—but are we really still part of nature's family? Or have we outgrown nature? Before we dive further into the philosophy behind our relationship with nature, we should be aware of another recent scientific discovery: We actually carry a lot of nature around with us all the time.

You Are Never Alone

In addition to our biological ancestry, the need to stop our suicidal war with nature, and the fact that a close relationship with nature would enrich all of us emotionally, economically, and spiritually, there is an additional reason why we should rethink our relationship with nature. That reason is that it is surprisingly difficult to define where the human body ends and where the rest of nature begins. The Human Microbiome Project, a collaboration between the National Institutes of Health in the United States, has for the first time systematically mapped the ecosystem that comprises our human body. In June 2012, the results of the five-year research project involving over 200 scientists from 80 universities and scientific institutions, were published in the journal *Nature* and several other scientific periodicals in the Public Library of Science (PLoS). The results are fascinating. The human body is home to trillions of microorganisms, collectively known as the human *microbiota*.[1] These microorganisms include bacteria, fungi, and viruses. They also consist of archaea and protists, which are simpler predecessors of bacteria.

Together, our microscopic passengers and symbiotic helpers can outnumber human cells at a ratio of 10 to 1. Because most of them are so small, they do not outweigh our human cells. However, our microbiota can still account for a respectable 1 to 3 percent of our adult human body mass, reaching up to 5 pounds (2.4 kg) in an adult weighing 176 pounds (80 kg). Our gut contains the most significant portion of our microbiota. All humans have their own composition of intestinal microbiota, like an internal ecosystem fingerprint, which is formed during infancy, mainly transferred through skin contact with our parents and ingested with our mother's milk.[2] The diversity of our internal ecosystem is essential for human health, including for our digestion, immune system regulation, and protection against pathogens.[3,4] We could simply not live without this large amount of "nature" within our own bodies and on our skin. We might have renounced nature as our true family, but nature has never renounced us.

Are these trillions of microscopic passengers and symbiotic helpers part of our human body, or are they part of nature? Where exactly would we draw the line if we are indeed separate from and external to nature, or even "opposed to nature," as modern dictionaries would have us believe? As we have discussed, drawing neat borders and boundaries that separate one thing from another in nature can be counterproductive. Like any mental model or map of the world, an image of nature in our minds cannot fully represent reality. All parts of nature are interconnected and interdependent at all times. Why, then, does it seem so difficult for modern Western culture to acknowledge that we are part of nature?

Nature Redefined

In most Western languages, the term "nature" refers to the living world, comprising organisms, ecosystems, and the biosphere. Most current definitions of nature explicitly or implicitly exclude humans; some do include us, particularly definitions from earlier than the 20th century and some newer definitions that have emerged in the 21st century. At the time of writing, the definitions in all modern English dictionaries still set nature and culture squarely apart. *Merriam-Webster's* dictionary defines nature as "the external world in its entirety," meaning external to humans. One of the nine definitions is "a creative and controlling force in the universe," which, in my view, is closer to the truth.[5]

The *Oxford English Dictionary* sets nature and culture even more squarely apart. Nature is "the phenomena of the physical world collectively; especially plants, animals, and other features and products of the Earth itself, as opposed to humans and human creations."[6] So, according to our use of the English language, not only are we *not* a part of nature, we are even "opposed to" nature. Why our conventional wisdom and use of everyday language places us and our human culture and civilization so clearly outside of nature, and how this affects our thinking and behavior, merits some attention. For argument's sake, let's assume that we and all human culture and technology are indeed external to nature. Where exactly is our place in relation to nature? External, perhaps, but are we above, below, or next to nature? Let's examine each of these theoretical options briefly in turn and determine how tenable or untenable these concepts are in practice.

Are We Above Nature?

As we see in the next examples, contradictory views on humanity's position in relation to the rest of life on Earth are currently widely held in parallel: According to different authorities, cultures, and contexts, humanity is at times perceived as above nature, below it, or next to yet separate from it. Some cultures still see us as an integral part of nature. But how can we simultaneously be above, below, next to, and inside something? The question of where exactly we position ourselves and our culture in relation to nature matters a great deal. Let's remember that science is telling us that we are rushing headlong into an extinction crisis of geological magnitude and of our own making, with as many as 1 million species at risk of disappearing.[7] Nature is crumbling all around us, and the impacts of ecosystem collapse are becoming increasingly apparent. When something large is about to fall, we want to be sure to know where we stand in relation to it and what we can do to avoid damage to ourselves from the crash. If collapses of critical ecosystem functions at a global scale or local scales indeed occur, questions will be raised about who or what is responsible for the loss of biodiversity and the resulting loss of human life and property. How much did we know, and why did we not prevent it? Therefore, it matters where we stand vis-à-vis nature, both individually and collectively.

Let's start with the view that humans are above nature, a belief that is probably the most commonly held regarding the position of *Homo sapiens*. The view is often implied and rarely openly stated, nor openly discussed. It is evident, for example, from the definition of nature in our legal system and from our behavior toward farmed animals and our treatment of nature as a commodity in our market economy. Our perceived superiority to nature is so fundamental to our current predominant worldview and the functioning of our current extractive world economy that questioning the basic assumption that we are above nature might seem radical. But where did our apparently deeply held assumption that we are above nature come from?

One possible explanation lies in religious worldviews that have shaped much of the modern world through centuries of colonialism and the dogma of humans being the sole center of the universe. The Bible states in Genesis 1:26–31: "God said, Let us make man in our image, after our likeness: and let them have dominion over the fish of the sea, and over the fowl of the air, and over the cattle, and over all the earth, and over every creeping thing that creepeth upon the earth."[8] This passage provides two reasons why we are superior to nature: The first is that we are created in the image of God, so we are distinct from other animals. The other is that God has given us dominion over all other life on Earth and, thus, over nature. How can we be part of that over which we hold dominion? This interpretation has dominated the Western view of nature and our role in the world for much of the past two millennia, perhaps even more so than the story in Genesis, where Adam and Eve were expelled from the Garden of Eden. However, much of Christian doctrine also encourages harmony between humanity and nature, and a more benign and modern interpretation of biblical guidance is emerging, as we see later in this chapter.

Other religions also have an open door for harmony with nature. The Quran has more than 750 verses that are related to nature and several passages that caution against overconsumption, stressing the stewardship role of humans in caring for and living in harmony with Allah's creation.[9,10] Since the significant shift in geopolitical power from the East to the West, which occurred gradually over several centuries but became most pronounced during the 19th and early 20th centuries with the Industrial Revolution, Western worldviews have played a fundamental role in shaping the rules of the modern world economy. Such rules developed mostly after the end of World War II, with the establishment of institutions such as the World Bank and the International Monetary Fund. However, religion alone cannot explain our perceived separation from nature. The main cause lies in recent European philosophy.

The view of absolute human dominance over nature was deeply engraved in the Western collective psyche in the 17th century, when philosopher René Descartes and others promoted a rationalist worldview that increasingly framed the universe in mechanical terms, likening it to a vast machine.[11] His perspective became central to Western philosophy during the so-called Era of "Enlightenment," the rather hubristic name for the period in history from the late 17th to early 19th centuries. Its most influential figures—such as Voltaire, Montesquieu, Rousseau, Diderot, Hume, Kant, and Adam Smith—were all European. Descartes' philosophy published in the mid-17th century and similar texts from other thinkers over the following decades rejected the idea of a divine presence in nature and all beings. Instead, it emphasized a sharp division between the human mind, with its capacity for reason and the seat of a divine connection, and the rest of the material world. Descartes' other enduring legacy is his concept of duality, which posits that body and mind are two distinct entities. This Cartesian dualism gives rise to the famous mind-body problem, which questions how two fundamentally different substances can interact. How can we be one entity when our mind and body are separate? If the mind or the soul can exist without the body (a fundamental teaching of all world religions), is our human identity more linked to the mind, or the soul, or the body? And if we are more the mind or soul, then is it not irrelevant what happens to the body and, consequently, to everything else in the material world, including nature and Planet Earth? The seeming disregard for the rest of creation in our Western worldview comes at least in part from viewing our bodies as the "weak flesh," to be disregarded and mastered by our will.

The mind-body problem is a challenging part of overcoming our perceived separation from nature. Like billions of others, I believe we are immortal souls, temporarily inhabiting a body, a commonly held belief across world religions. However, I also firmly believe that this concept of our dual existence as both a body and a soul gives us enormous responsibility for our fellow human beings and all life while we live on Earth. It is easy to see, though, that the mind-body problem might serve as an excuse to disregard anything material as less essential and less worthy of our regard than the spiritual realm. It was, therefore, a historic development when Pope Francis published his encyclical *Laudato Si'* in 2015 to correct this misperception, at least for Catholics. We explore the importance of the pope's encyclical as a new Christian doctrine later in this chapter.

The Clouds of "Enlightenment"

Other philosophers of the 17th and 18th centuries supported Descartes' distinction and hierarchy between humans and the rest of nature. For instance, Thomas Hobbes argued that all animals operated purely on instinct, without reason or feelings.[12] Baruch Spinoza viewed human reason as unique in its ability to grasp the laws of nature, which he believed governed everything else.[13]

Immanuel Kant later built on these ideas, stressing that humans alone possess moral autonomy, separating them fundamentally from animals and the natural world.[14]

It is clear that animals do not have the same ability as we do to communicate, reason, and plan—although the quality of our human abilities to reason and make wise decisions varies widely across individuals and social groups. Despite our unique gifts, we do not always make wise choices. The ideas of Hobbes, Spinoza, Kant, and many other thought leaders, including the Christian Church, reflected scientific knowledge in biology and ecology of the time. Animals were poorly understood, and the intricate connections between species in an ecosystem were not yet widely known. However, since then, we have overwhelming scientific evidence that animals think, can reason, and feel deeply, possibly across almost the entire spectrum of human experiences. Animals use tools, make informed choices based on collecting information, and can form lifelong bonds—not just with their own species but with other animals and humans as well. In his book *Beyond Words: How Animals Think and Feel*, Carl Safina describes the latest scientific discoveries, including the fact that elephants call each other by name, as do dolphins and probably whales.[15] Why would animals need names for each other if all they ever do is act on instinct, like preprogrammed machines? And if this basic assumption of the "Enlightenment" thinkers is wrong, perhaps other aspects of that philosophical framework are also outdated.

Based loyally on our "autopilot" of religion and philosophy, current economic textbooks still portray nature mainly as a commodity. In many countries, our relationship with nature is governed by ministries or departments of "Natural Resources." Although there is a growing recognition of nature's importance beyond economics, the depiction still often leans toward viewing nature primarily as a resource for economic activity, most recently referred to as "natural capital." We dive more into the relationship between nature and our economy in Chapter 4. Traditional Christian religion, as well as most Western philosophy, clearly positions us above and beyond nature. The problems caused by this self-serving worldview, notably destructive pressure on other species and on ecosystems, have given rise to an opposing school of thought that places nature above humanity, akin to a deity. However, as we will see, that view of nature is not useful either.

All Hail Nature, the Divine?

About 200 years ago, a strong tendency emerged in Western literature to romanticize nature, with Henry David Thoreau's seminal work, *Walden,* being perhaps the best-known example of deifying nature and criticizing the human role in shaping, using, and managing her.[16] My most striking encounter with this romanticized view of nature was a few years ago, when my then 10-year-old daughter described humans to me as a disease inflicting Planet Earth and spoiling its natural beauty. My daughter has since then softened her harsh stance on our species, but at the time, she saw nature as a supreme entity deserving to be left in peace by humans. That view is widely held among conservationists, but it is misleading. Humans are not a disease on the rest of creation. We have the ability to manage nature sustainably and even heal and regenerate nature in a symbiotic relationship. Modern conservation organizations are quick to assure us that they "put people first," in particular, Indigenous and local communities. However, beneath the surface lies a lingering perception among conservationists that we must protect nature from humans. In fact, nature can never be effectively protected *from* people, but only for and by people.

In my experience, those who are most difficult to convince that we are part of nature are often nature lovers, including conservationists, biologists, botanists, and enthusiasts of the outdoors in general.

I have observed a certain degree of misanthropy in these circles, to which I also belong. In my youth, right after high school, I worked for one year as a national park ranger on the northwestern coast of Germany, on the island of Sylt, famed for its long, sandy beaches. The island lies in the Wadden Sea, the large marine tidal estuary of the North Sea, which is one of the world's most important migratory bird habitats. At the time, the area was a relatively young national park, having been established only a few years earlier. There was still considerable conflict between the local population, who grudgingly complied with the numerous new rules and restrictions that suddenly governed their landscape and seascape, including restricted access to certain beaches. We spent a lot of time discussing and explaining to locals and tourists how protecting nature would ultimately benefit local communities through increased tourism and improved fishing grounds around the marine protected areas.

During a recent guided walk with former volunteers, some of them now employed by the National Park Service, we discussed that the dunes on the island were overgrown with an invasive species that was suffocating the natural flora. The new recommendation was to allow the estimated one million tourists who visit the island each year to roam freely across the dunes, thereby trampling and opening up the areas for the heather and other native plants that once grew there. When I suggested that this confirms that humans are part of nature and that we are ecosystem engineers like many other species, my fellow conservationists were dismissive of the idea.

Their negative response surprised me until I understood that the same argument—that humans are part of nature—is frequently being used, and perhaps misused, by the coastal fisherfolk, including proponents of the profitable blue mussel fishery that harvests millions of mussels each year from the national park for export. They argue that because humans are part of nature, they should still have the right to fish and cultivate mussels even in the core area of the park. In my opinion, this debate exemplifies how humans are an integral part of nature. However, it is also an example of how easily we can overstep our limits as a species. Releasing and growing large numbers of blue mussels on artificial reefs near the shore is a complex aquaculture operation. Despite best efforts, some damage to the seafloor and its fragile habitat occurs when mussels are harvested, which is why conservationists want to ban mussel fisheries altogether from the park.

Yet the mussel fishing industry provides employment and income for the island, and the artificial mussel banks serve as a habitat for many marine species. If humans are part of nature, national parks should be there for us as much as for other species, including for recreation and to benefit from the ecosystem. In the case of coastal fisheries and aquaculture for blue mussels, we can ensure that this production occurs with benefits for both humans and nature. It is possible to plan and harvest mussel banks in a sustainable manner. The question is how we behave and what our relationship with nature is—in national parks and everywhere else. The relationship should be and can be mutually beneficial rather than a one-sided or toxic one that will inevitably collapse over time. In this debate, as in many others around the world where we are coming into conflict with nature, our philosophy and our intention matter.

Counterproductive Conservation

When I speak of misanthropy in conservation circles, I do so with all due respect to the many women, men, and children who have defended nature, often at great personal risk, and sometimes with their lives. What I mean by *misanthropy* is that conservationists can sometimes feel a sense of despair when it comes to the widespread greed, apathy, and lack of basic respect for nature among many people in power and the general public that leads to our current nature crisis.

That despair can lead to the conclusion that most humans are ignorant and indifferent to the beauty of nature. How could such people ever be part of nature? And if those people are not part of nature, then clearly none of us can be considered part of nature, since we are one species. Following this line of thought, it is better to maintain the belief that nature is separate from us and elevate nature as both external and superior to humanity. I believe that the deification of nature is no more helpful or constructive than believing ourselves superior to nature. Whether we perceive ourselves as above or below nature, both concepts result in a mental and emotional distance that is counterproductive to a healthy, long-term relationship.

A 30 Percent World

A world where both humans and nature live in harmony and thrive will be possible once we change the core of our relationship with nature. In 2022, governments around the world agreed on an ambitious plan to protect 30 percent of the world's land and seas by the year 2030.[17] The so-called 30 by 30 plan was widely hailed as a success for nature. However, let us examine briefly what this implies for the other 70 percent and for our relationship with nature. Seen from the perspective of a long-term, healthy relationship, aiming to protect 30 percent of the world's land and sea from ourselves seems more like a declaration of moral failure rather than a decisive victory. At most, it is a desperate emergency measure that might slow the decline of nature but still might backfire in the long run. Because the 30 by 30 plan was so widely celebrated as a decisive success for nature, decision makers now believe we have won the war in our relationship with nature. In reality, the 30 by 30 plan might achieve some conservation success, but it also risks further solidifying the predominant worldview that it is inevitable and justifiable that we will continue to degrade and pollute the other 70 percent of Earth.

Of course, there are also policies and efforts underway to take better care of the remaining 70 percent. The Global Biodiversity Framework adopted in 2022 by the Convention on Biological Diversity has a laudable vision of humanity living in harmony with nature by 2050 as well as ambitious targets for the sustainable use and restoration of nature. However, when it comes to making important financial and legal decisions, the lowest common denominator tends to prevail when negotiating issues at a tactical level in politics rather than discussing societal values, such as human rights. In other words, we need to elevate our relationship with nature to a strategic political and philosophical level, backed by a new societal understanding that we are an integral part of nature. Once nature is understood and viewed in this light, it becomes clear that we cannot ignore nature outside of the 30 percent of protected areas (and most countries are still well below that level, by the way). Nature should be at the forefront of our minds in all key decisions regarding land use, forestry, fisheries, water consumption, and critical sectors such as energy, water, and food. Plans like the 30 by 30 are well meaning but ultimately insufficient. Worse, they can give policymakers the illusion that if we protect just 30 percent of our land and sea, the "nature problem" is taken care of, and we can go back to business as usual.

Let's try to see the 30 by 30 campaign from the standpoint of a politician who has finally started to look into this "next big thing" called nature and has heard and heeded the persistent call from biodiversity experts and the public that we need to save biodiversity from collapse. What is the loudest, simplest-sounding solution? Setting aside 30 percent of protected areas for nature globally.[18] By yielding to the campaign and allocating some public funds to the 30 by 30 goal and to biodiversity interest groups, politicians have addressed this specific problem called nature. Or so they might believe.

Let me be clear. More space for other species and more protected areas are promising developments, but they will not succeed unless humans can also directly benefit from those areas. We are part of nature. And the 30 by 30 plan is only useful if we acknowledge that what happens on the other 70 percent of Earth's land and sea is even more important. Thirty percent might not even be enough to safeguard the most critical ecosystems or critically endangered species, because protected areas tend to have static boundaries while climate change is already causing almost all species to move. More importantly, it will not resolve our relationship crisis with nature but keep us in the status quo of "we are over here, locked into business as usual, and whatever is left of nature is over there, locked away from us." If not accompanied by a new narrative about our relationship with nature, a narrative of protection might increase the mental and emotional distance between us and nature even further.

More and more frequently, high fences separate humans and the rest of nature—in many cases, electric fences. In Africa, this measure is often a necessary last resort for conservation to avoid poaching and to keep large mammals like rhinos from escaping their enclosed and guarded habitat. But the fencing of nature is spreading. Recently, in my country of residence, the Danish government has allocated significant amounts of land as new national parks, some of which will be heavily fenced to keep in reintroduced large herbivores, such as moose and horses, which are brought in to mimic age-old ecosystem processes. Only a select few humans can experience the richness of nature in these newly protected areas. Scientists can study nature even in "Category I" protected areas, the most strongly protected category defined by the International Union for the Conservation of Nature. In contrast, the areas are off-limits for the rest of us. What does that say about our relationship with nature? From the reactions of my neighbors in our rural town, so far the measure has not contributed to healing our relationship with nature.

Ultra-High-Net-Worth Nature

Another example of a misguided approach to saving nature by separating ourselves from her is the increasing tendency of the super-rich to buy up land in the name of conservation or restoration. Although they might indeed have altruistic motives, their land purchases often just increase their enjoyment of nature by excluding everyone else. Ultra-high-net-worth individuals are increasingly acquiring extensive natural estates and entire landscapes. They often do this for hunting, wildlife viewing, or simply as a good investment during times of plummeting wildlife populations, when intact nature is becoming a scarce good. A public good that is becoming rarer by the day will increase in price when privatized. For example, in Kenya, private nature reserves are now frequently home to multimillion-dollar mansions and other investments by high-net-worth individuals. It doesn't feel right that intact nature is becoming something only the very wealthy can enjoy.

Instead of focusing all our efforts on protecting 30 percent of nature by 2030, we need a much more ambitious plan, based on a philosophical and socioeconomic reset in our relationship with nature. We need a 100 percent plan for nature and humans to live in harmony. And it has to start right now, instead of deferring the goal of humanity living in harmony with nature to 2050, as specified in the Global Biodiversity Framework. We need to transition to 100 percent of nature under sustainable management, including areas for conservation and restoration, with the majority under sustainable agriculture, fisheries, or forestry practices. As long as our long-term goal is to start living in harmony with nature, we can view the 30 by 30 plan as a first step. Let's remember, however, that there is a more fundamental issue to fix. We want to live in a 100 percent healthy, abundant, and diverse world. We need to fix our entire relationship with nature, not just one-third of it, and make that relationship more reciprocal and dynamic instead of more exclusive and static. Fortunately, a broad societal understanding that we are part of nature is now taking hold in mainstream public discourse.

A New Hope

The "Enlightenment" Era is several centuries behind us and all its foremost writers were white European males from a similar social class. Nevertheless, it remains a defining thought framework underlying and guiding our Western societies and the world economy. And although we still are in an undeclared "suicidal war with nature," as UN Secretary-General António Guterres put it, there has been some progress toward making peace with nature. Just a few decades ago, an open war between humanity and nature existed. During his rule in China from 1949 to 1976, Mao Zedong established a "war on nature"—a series of campaigns and policies that aimed to dominate and reshape the natural environment for human benefit, often with disastrous consequences. This approach was summarized in Mao's slogan "Man must conquer nature," representing a departure from traditional Chinese philosophy, which emphasizes harmony between humans and nature.

Another example of open war on nature in our recent past hits closer to home for me. Before Rachel Carson's book *Silent Spring* was published in 1962, the pesticide DDT was in wide use worldwide to kill all insect life in its path. My mother recalls planes spraying DDT in large white plumes over the forest next to their house in the 1950s to fight a local bark beetle outbreak. Carpet bombing a forest ecosystem with a highly toxic organochloride that accumulates in the food web, including in humans, is a thing of the past, one hopes.

We have also moved beyond the time when animals were seen as "things" in modern science. When Jane Goodall submitted the first manuscript of her groundbreaking behavioral research on chimpanzees to the editors, they requested her to change all the pronouns of chimpanzees from "she" or "he" to "it." She refused and got the manuscript published anyway, winning a small but significant victory for all nonhuman species. We have come a long way since then, thanks to Jane and other pioneers like her who see more in our relationship with nature than that of a consumer and a commodity or that of a ruler who dominates the natural kingdom. It is time to take the next step and reaffirm our close relationship with nature.

We are on the promising track but still have a long way to go. The perception that nature is external to us has even triggered the creation of a new word: the "environment." Derived from the French *environs* (meaning "around us"), this term was hardly in use before the year 1800, but around 1900, the "environment" began its steady climb in widespread use, mirroring the decline of the word "nature" in literature and media, until "environment" almost overtook "nature" as a term to describe the natural world.[19] But in the year 2000, that tide turned. When analyzing the frequency of both terms in literature from 1800 to today using Google's Ngram Viewer, "nature" has been on the rise again in recent decades. This is a sign of an emerging evolutionary step in our collective consciousness, indicating that our voluntary exile from nature is coming to an end. Nature is not the environment. Nature is us. We know it's time to come home.

Coming Home

The "Enlightenment" Era philosophy, the Atomic Age, and, more recently, the power to create large language models and algorithms that are much more intelligent than humans (at least if we measure intelligence in computing speed, pattern recognition, and reasoning ability) have brought revolutionary developments in art, philosophy, technology, and politics. However, our technical progress has also tempted us into enormous hubris and has distorted our view of all other species as distinctly and decidedly inferior to us. This superiority and dominance approach has served and is still serving as an excuse for an industrial-scale onslaught on many nonhuman species. If they are just intricate machines and only we are blessed with divine intelligence, what stops us from seeing animals, plants, and entire ecosystems as soulless commodities put here for our convenience? Our perceived separation from nature allows us to view them as mere things to be taken, eaten, burned, and liquefied into financial capital without a second thought and without any reciprocity or any relationship between us and them.

In our generation, we must move beyond the worldview that we are above nature. We have hollowed out nature to such a degree now that if we really sit above it, we are about to take a deep fall. However, the most important reason for resetting our relationship with nature is that we are missing out on a beautiful relationship of mutual respect, a relationship that would enhance and enrich our lives significantly. It is time to set things right. Fortunately, some of the leading thinkers of our time agree.

In 2015, a radical and powerful publication was written to destroy the myth that we are above nature. The head of the Catholic Church, Pope Francis, published his seminal work, *Laudato Si'*, in which he clearly states that "nature cannot be regarded as something separate from ourselves or as a mere setting in which we live. We are part of nature, included in it and thus in constant interaction with it" (para. 139).[20] The text was addressed to the world's 1 billion Catholics and all of humanity. Pope Francis also acknowledged that "never have we so hurt and mistreated our common home as we have in the last two hundred years" (para. 53) and that "the harmony between the Creator, humanity and creation as a whole was disrupted by our presuming to take the place of God and refusing to acknowledge our creaturely limitations" (para. 66).

It is difficult to overstate the importance of this reversal of Christian doctrine. Pope Francis—who chose his name after Francis of Assisi, probably the most radical and humble of all saints—wanted to be sure that this new view received wide attention, so he wrote this encyclical in Italian instead of the traditional Latin, the first time a pope had done so. The pope clearly and forcefully articulated the new paradigm that we are part of nature. Writings such as *Laudato Si'* are starting to change our worldview, shifting from the notion that humans are above nature to one that recognizes humans as a part of nature. Other world religions had recognized the truth about our relationship with nature sooner. In Buddhism, for example, the relationship between humans and nature is seen as deeply interconnected and symbiotic. At its heart is the understanding that all life-forms, including humans, are part of a larger, interdependent web. Nature is not separate or "out there"; we are inherently and inextricably part of it.

Words Matter

Increasingly, writers, philosophers, and thought leaders are questioning traditional dogmas and creating a new global narrative that acknowledges that we are indeed part of nature. My favorite indication that things are changing comes from the Secretary-General of the United Nations, António Guterres, who has spoken frequently about our need to stop our suicidal war with nature. Based on the findings of the UN's Intergovernmental Platform on Biodiversity and Ecosystem Services, we are on the brink of a mass extinction crisis. In his "The State of the Planet" address at Columbia University in 2020, the Secretary-General pointed out that "making peace with nature is the defining task of the 21st century. It must be the top, top priority for everyone, everywhere."[21] The UN Decade on Ecosystem Restoration 2021–2030 puts repairing our relationship with nature at the heart of its strategy, with its central vision being "a world where—for the health and wellbeing of all life on Earth and that of future generations—the relationship between humans and nature has been restored."[22]

Scientists are also coming to believe that we need to reposition ourselves in our relationship with nature. Tom Oliver is a professor of applied ecology at the University of Reading and cofounder of the We Are Nature campaign that lobbies dictionaries to change their definition of nature and include humans in it. He sums it up aptly:

> *[T]o change the primary definition of nature from "as opposed to humans" to "including humans" will require more people to use the word in a way that reflects how humans are intertwined with the whole web of life.*
>
> *The great thing is, by doing this, we rekindle the bonds of care towards the living world around us. And by dispelling the illusion of our separation from nature, we can also expect to lead happier lives. Words matter—there is restoration and joy from talking about how we are nature.[23]*

Our thoughts and words give form to our relationships. Having the right basic vocabulary allows us to freely express a newfound relationship with nature in our individual way. Our relationship with nature is unique for each person on Earth, but there are also commonalities. It would be highly beneficial if we, as a society and global community, could find as much common ground as possible when we define humanity's relationship with nature. So, let us recognize the fact that our dominant Western mental model of the world is several centuries old, dating back to the "Enlightenment" Era. Perhaps it is time to give that classic worldview a complete overhaul.

Upgrading the Western Worldview

None of the science about the human microbiota, quantum theory, astrobiology, the DNA we share with other species, or the fact that humanity would ever reach a point where nature was on the verge of collapse at a planetary scale was known to European philosophers in the 17th and 18th centuries. Nor to the men gathered for the Synod of Hippo (393 CE) and the Councils of Carthage (397 and 419 CE), who affirmed the 27 books of the New Testament and essentially compiled the Bible in its current form. It is time to reconsider our isolated and somewhat aloof and philosophically and scientifically untenable position that we are the only divine and intelligent beings among millions of other life-forms on Earth. The worldview of dominion, oppression, and the extraction of nature as a commodity has reached a dead end, both philosophically and practically.

The fact that we are indeed part of a larger family of animals and plants, and thus part of nature, has always been known to many of the world's Indigenous peoples. Indigenous activist and author Nemonte Nenquimo sums it up aptly: "I think that we are all connected to the Earth, but some people have been spiritually disconnected."[24] She explains that for her people, the Waorani in the Ecuadorian Amazon, there is a "deep spiritual connection with Mother Nature and how we interact as people, in a very close way with the Earth, with animals, and with all that is sacred."[25] In her book *We Will Be Jaguars*, Nenquimo details how this connection clashes with the vastly different predominant Western worldview of resource extraction, colonialism, and dominance over nature and Indigenous peoples.[26]

As we see in Chapter 5, a worldview of kinship and a reciprocal relationship with nature, rather than one of dominance, leads to very different behaviors and outcomes. The traditional ecological knowledge of Indigenous peoples and local communities around the world has much to teach modern societies, starting with the fundamental question of who we are to ourselves, to each other, and to nature.

So, Are We Part of Nature?

With the powerful knowledge and tools that we have gained over the past centuries, we have grown up to be not just another member of nature's large family. We are now the world's most powerful ecosystem engineer, with the power to destroy, transform, or restore nature on a planetary scale. With that power comes enormous responsibility. We are responsible for stewarding our natural heritage back to a level of natural wealth where humans and all our fellow species on this planet can have a life worth living, in perpetuity. This is our birthright, and it is the birthright of all creation. Nature is our home and our family, and that family needs us now as much as we have ever needed them.

A generational awakening is happening across our planet as we realize that our interdependence with nature runs deep and that our separation from nature serves neither us nor any other species. The recognition that we are part of nature will start a much-needed healing process for nature and humankind. Healing and restoration can occur in both directions. It feels good and right to acknowledge our place among nature as part of our family, to take our proper place, and to accept our responsibility. Acknowledging that we are part of nature is a liberating and spiritually uplifting thought. We are part of all life, and all life is part of us.

We all originate from the same physical source—energy bound as matter—and are connected by the same life force. Whatever our faith or creed, it probably already tells us to care for creation. Let us take that more literally, as if we are asked to care for our neighbors.

The scientific facts are convincing enough: We are part of nature, and nature is one single connected system across the biosphere of Planet Earth. When even the Catholic Church, not known as an early adopter of groundbreaking science, declares that we are part of nature, it is probably time to put 300 years of the reductionist thinking of "Enlightenment" philosophy behind us and embrace the fact that Earth is a living system with millions of interconnected species. We are part of this system, not above it, outside of it, or opposed to it, and we should live our lives accordingly. That recognition also means we can collectively rise to become nature's champions and restore ecosystems at a large scale. We need a new everyday definition of the word "nature." Nature is the entire part of the universe that contains life, including humans. It's time to update our dictionaries and to update and open our minds. Once our minds and hearts are open, a new relationship with nature becomes possible, even inevitable. We are part of nature and we can do better to give back. In the next chapter, let's look at how we can assess, value, and account for nature better in our global economy. And recognize why every job is now a nature job.

Chapter 4

The Value of Nature

Nature is our home. Good economics demands we manage it better.

—Sir Professor Partha Dasgupta

In 2020, when my family and I purchased our small farm in the Scandinavian countryside, there were only a few financial incentives for farmers to adopt regenerative agriculture practices. However, over the past few years, several new subsidies have been introduced across the European Union to support regenerative farming, a practice that restores the health of the soil. Recently, no less than three different organizations offered to pay our family for planting a native forest on one of our fields. We went with a new public–private partnership, the Climate Forest Fund, to afforest about 20 acres (8 ha) of cropland with 26 different native species of trees and shrubs.[1] The government-sponsored fund, supplemented by additional contributions from the Danish State railway and other companies, covered all costs associated with the trees, planting, and fencing. Nature is starting to get the economic backing we need. As we see in this chapter, a global transition is underway to account for nature's true value, and we are close to a social tipping point where mainstream economics will be based on "the understanding that we—and our economies—are 'embedded' within Nature, not external to it."[2]

Just as we had the opportunity to build nature into our family farming plans in Denmark, many countries are providing support mechanisms for nature investments, including public incentives in Australia, Brazil, Canada, Japan, the United Kingdom, and the United States. Importantly, many young people are discovering a passion for working with land and soil in a holistic way. The coming generation's interest in regenerative farming is crucial if we want to keep agriculture, with its aging workforce, alive and thriving as an economic sector around the world. Similar sustainability transformations are occurring in other sectors worldwide. Fortunately, nature and her many functions are starting to be reckoned with in economic decision-making. It is about time.

"It's the Ecology, Stupid!"

The quip "It's the economy, stupid!" by James Carville during Bill Clinton's 1992 presidential campaign also rings true for efforts to restore nature at planetary scale, though with a twist. Economy and ecology have the same root in the ancient Greek word *oikos*—our common home. *Ecology* is the knowledge of our common home, and *economy* is its management. Managing our common home, Planet Earth, without knowledge about its inner workings is like flying blind. Ecological principles should inform how we structure and steer our economy. A new relationship with nature, in which we recognize that we are part of it, will have far-reaching consequences across our economy and culture. The most urgent and vital change is in how we plan and manage our economic activities, including production, purchasing, selling, and investing in goods and services. All of these activities depend to some degree on nature or, more specifically, on the goods and services produced by the complex web of life known as biodiversity.

Better data on the role of nature in our economy and better integration of this data are starting to change economic behavior, regardless of political ideology. Relying on the best available, long-term financial information and models to make sound economic decisions has nothing to do with being "woke." However, even when we have the best available information, societies themselves must choose who should pay for what in our society and economy. How much of the costs of conserving and restoring nature will be shifted toward all taxpayers rather than toward the companies that benefit from nature's goods and services? And how much of these costs are we prepared to shift to future generations?

Unlike physics, economics is not a natural science with predetermined natural laws. Economics is a social science, and the rules that govern it are social constructs, which we should update regularly to ensure they are fit for the future. Fortunately, most of the groundwork for a required fundamental upgrade of our economy to better account for nature has been laid. Today's decision makers can build on a rapidly growing body of knowledge on the economics of ecosystems and biodiversity. With some additional awareness raising, advocacy, and capacity-building efforts targeting central banks, treasury departments, and the finance industry, the global economy can begin to transition from an extractive and ultimately self-destructive model to a regenerative and circular one, which benefits both present and future generations. This shift is already well underway, though it is mostly hidden from public view—not because this work is secretive, but because it can be technical and at times tedious. Have you ever heard of the United Nations Statistics Division and its *System of National Accounts*, for example? Probably not, yet the regularly updated *Handbook on Management and Organization of National Statistical Systems* is one of the most important global documents guiding how governments make decisions, record progress, and measure national wealth.[3] "Natural capital," which is usually defined as the world's stock of natural resources including geology, soils, air, water, and all living organisms, was introduced into the official UN statistics in 2012 as a comprehensive new framework called the "System of Environmental-Economic Accounting (SEEA)."[4] Governments around the world are now starting to measure natural capital systematically, and national finance authorities are accounting for nature in their countries' balance sheets. It has taken many years to get to this point; progress at the UN level can be slow, but it is steadily and quietly transforming how nature shows up in our economy. It may seem obvious that if a country pollutes its water, cuts down all its forests, and overfishes its marine areas for short-term financial gain, it does not increase its national wealth. It just exchanges one form of capital (nature) for another (money) in a short-term transaction, which makes everyone worse off in the long term. However, if we look only at gross domestic product (GDP) as our main indicator of national wealth, countries that deplete their natural capital and gain financial capital can appear to do well, at least on paper.

The True Wealth of Nations

Increasing the true wealth of a nation means growing the well-being of its people, as well as its social and natural capital, simultaneously with its financial capital. Some countries have achieved this. For instance, Costa Rica has more than doubled its forest cover and increased its GDP tenfold in recent decades; we explore this further later in the section "The Cost to Restore Earth." Omitting the true value of nature in national accounting, a fundamental error in economics that has persisted for hundreds of years, is being fixed now. Since the introduction of the UN "System of Environmental-Economic Accounting," countries have begun to measure systematically whether natural capital is being depleted or replenished. Because we are becoming better at measuring what we want to manage, we need a more holistic and accurate measure of the wealth of nations and societies.

Change is afoot in other chambers at the heart of the economy as well. Let's take as an example the United Kingdom's Institute and Faculty of Actuaries. If you don't know what actuaries do, we are in the same boat—I only found out recently myself. They are the experts who compile and analyze statistics and use them to calculate our insurance risks and premiums. That sounds like a relatively niche job, but it is a central function in our economy, as it determines the cost of everything (at least if you want to have it insured). Even the rather conservative and opaque profession of actuaries is starting to include nature and climate risks and rewards into their calculations.[5] Due to updated prices of these risks, you can expect the prices of your health, home, and life insurance to change soon, based on the level of access to nature and green spaces you and your community have. Long-term ecological risk, like building your home in a flood-prone wetland, will also be priced more accurately based on nature's ability to sustain a stable climate and other essential goods and services for humanity.

These and other ongoing shifts in our economic framework are building on landmark assessments like the Economics of Ecosystems and Biodiversity, a concerted international effort between 2007 and 2011 to review the impacts of key sectors of the economy on nature and make nature's values visible, led by Indian economist and banker-turned-environmentalist Pavan Sukhdev. What sounded like an outlier idea in the 1990s—that we need to change our economic textbooks to account for nature—is now becoming mainstream. In 2025, the UN published a report estimating that the global economy is currently losing 25 trillion USD per year, or about 25 percent of its GDP, because of nature loss.[6] In other words, we could all be 25 percent better off and live healthier, happier lives in abundant nature if we acted decisively. The next time you vote, pay taxes, or are otherwise in contact with your local or national government, remind them that we could all be better off with more nature.

89

The Value of Nature

The same goes for the companies and brands you buy from and any financial investments, such as your pension. Every voice, every vote, and every dollar counts.

The True Costs of Nature Loss

In our secular, utilitarian world, we tend to value and manage only what we measure. It is a good sign that governments have started to calculate and monitor the value of everything that comes from nature—"soil, air, water, and all living creatures"[7]—in a systematic way. Even if governments don't measure all stocks and flows of natural capital, it is becoming more common now for heads of state and ministers of finance to have insights into the natural wealth and natural capital of their nations. A lack of these insights can lead to detrimental policy decisions, such as the government-driven Indonesian "food estate program," which cleared vast areas of forest between the early 1990s and 2020 to plant rice, soy, sugarcane, and cassava. What appeared to be a good plan on paper ultimately proved disastrous. In addition to fueling climate change and biodiversity loss, one unintended outcome of this large-scale deforestation is that vast parts of Indonesia are now essentially lost to future generations. The soils under the cleared tropical forests were often peatlands—basically, a form of compost built up over hundreds of years and locked airtight underwater. When peat comes into contact with air, it starts to decompose, releasing most of its stored carbon back into the atmosphere. When peatlands are clear-cut, drained, and plowed for agriculture, they begin to subside: They shrivel up and contract as the organic matter dries up and decomposes. The entire landmass of Indonesia's coastal lowlands is now literally sinking into the sea, at speeds of up to 8 inches (20 cm) each year.[8] Saltwater starts to infiltrate the sinking coastal soil, making it infertile and toxic for most plants. Even the capital city of Indonesia, Jakarta, is slowly sinking into the sea, although the main cause of subsidence here is primarily excessive groundwater withdrawal for drinking water. The soils on many former Indonesian peatlands, such as in Central Kalimantan on the island of Borneo, have become too acidic to generate agricultural yields. Agricultural expansion and deforestation have created some short-term economic benefits for a small minority of shareholders, mostly from selling the tropical timber during the conversion of old-growth forest to agricultural land. However, the degradation of nature ultimately resulted in a significant socioeconomic loss for all Indonesians. If the cost-benefit calculations for the food estate program had included the actual costs of natural capital, the disaster might have been averted.

The Costs to Restore Earth

Turning the tide on nature loss is possible. Once we recognize that conserving and restoring nature requires more funding, it is helpful to view this as an investment rather than an expense. So, how much investment is needed? Let us use the most recent estimate from the United Nations. The Intergovernmental Science-Policy Platform on Biodiversity and Ecosystem Services (IPBES) estimates that we need around 1 trillion USD per year to halt and reverse nature loss globally.[9] Other estimates, such as the UN's State of Finance for Nature report, have a slightly lower estimate, but it is in the same ballpark.[10] One trillion dollars sounds like a lot of money. However, regaining natural capital at scale is an excellent investment considering it means that we can reclaim some of the 25 percent of GDP that we are losing annually due to nature loss. Seen from that perspective, we need to invest only a small percentage of the global economy, approximately 1 percent of our global GDP. This is much less than public spending on fossil fuel subsidies, estimated by the International Monetary Fund to be about 7 percent of GDP per year.[11] Investments in nature restoration can generate a societal return of up to 30 USD for every 1 USD and are among the highest job creation investments of any public or private spending.[12] An excellent rate of return, even for Wall Street investors. Most of the returns would accrue in the form of ecosystem services, such as clean water, which would benefit society at large rather than necessarily individual investors, but that fact makes the macroeconomic case all the more compelling.

One trillion USD is also not a lot of money compared to the amount governments and the private sector already invest globally each year in what we could call "nature-negative" investments: destructive or extractive economic activities that harm nature. The UN Environment Programme lists the amount of such harmful spending at 6.7 trillion USD per year—6,700 billion USD per year, or about 12 million USD every minute that your and my taxpayer money is spent on further degrading nature and our only home.[13]

Examples of such ecologically harmful incentives include fossil fuel subsidies and public subsidies for pesticides. In total, these nature-negative investments account for more than 5 percent of the entire global economic output per year. That is approximately as much as all public spending on human health. This harmful waste of taxpayer money can and will change as the hidden costs of nature loss become clear and the opportunities are better understood. Subsidies and tax incentives are the primary tools governments have to shift consumer and corporate behavior. Shifting the spending of public money from exacerbating the problem to healing and restoring nature will send strong long-term market signals, prompting a corresponding shift in private-sector investments. This shift is already happening in a few countries.

Costa Rica had one of the highest deforestation rates in the world in the 1980s, with forest cover falling from 72 percent in 1950 to 21 percent in 1987.[14] In 1997, the country began redistributing 3.5 percent of its national fossil fuel tax revenue to support landowners in the conservation and restoration of forests. This policy, combined with other incentives and income streams such as ecotourism, was so successful that Costa Rica's forest cover increased to over 51 percent today. At the same time, the country increased its economic output per capita 10-fold, from around 1,300 USD per person to 13,000 USD, laying to rest the myth that developing countries need to slash and burn through their natural capital as the only way to "develop."

Public subsidies to restore nature can be applied in other countries, but lessons about this successful approach have not yet been widely replicated. We need imagination to restore natural abundance and diversity at a national scale, and real-life examples are best way to create a clear, long-term vision of success. Large-scale restoration examples and a clear documentation of benefits will give politicians the courage and motivation to invest significantly in nature. And few things are more inspiring and motivating than success. Before turning to global success stories, however, let us further examine the role of nature within the engine room of our national economies.

A New Economy Emerges

The emerging understanding that humanity, including our economy, is part of nature received a significant boost in 2019 with the UK Treasury's commissioning of a landmark study. Led by Professor Sir Partha Dasgupta of the University of Cambridge and supported by an international advisory board of leading economists and other experts, the *Dasgupta Review* focused on the central question of understanding the economics of biodiversity and finding ways to incorporate nature into our everyday economic decisions.[15] The review, published in 2021, explores how biodiversity affects our national economy and overall human well-being. It demonstrates that current measures of economic success, such as GDP, do not accurately capture the true wealth of nations or the health of our economies.

The study proposes a new economic paradigm that treats nature as a vital resource we need to protect and that balances what we take from nature with what it can sustainably provide.

The *Dasgupta Review* helps policymakers, businesses, and society understand the real value of nature and why protecting biodiversity is crucial for our future. One can arguably boil the 600-page report down to a single phrase: "Pay for what you use." That sounds like common sense—and in most other aspects of our lives, common sense plays an important role. Economics, however, sometimes is far removed from common sense; in particular, when so-called externalities distort what things cost. Clean air and clean water, for example, were long considered free, public goods that did not require investments or safeguarding. They were considered "external" to economic decision-making at the individual or corporate level. Before the entry into force of landmark legislation in most countries, such as the Clean Air Act in the United States in 1970 or the EU Water Framework Directive in 2000, many economies treated their air and water like open sewer systems and did not invest sufficiently in keeping them clean and functioning properly.

Sooner or later, though, these hidden costs, or "externalities," have a way of catching up with the real price of a good or service. Usually, if a good is underpriced, it leads to market distortions, and someone ultimately will have to pay the full price. So it is with clean air and clean water: When they are polluted or run out, the public and taxpayers usually have to foot the bill. Internalizing the externalities into the economy means that those who use something have to pay for it, as the "polluter pays" principle states. That sounds logical and fair, yet unfortunately it is not how a market economy works, unless we regulate it sufficiently to ensure that companies that use a lot of water, for example, also invest sufficiently in the cleanup of wastewater and in the replenishment of water basins, through conservation or restoration of watersheds. Or they pay enough in taxes to enable the government to play this role. Suppose companies or individuals are allowed to use nature and her public goods for free and to degrade, pollute, or deplete a groundwater reservoir. In that case, their products may be inexpensive to produce, and shareholders might reap a profit that is disproportionate to the true cost of production, but the public ultimately bears the real production costs through taxes or damage to public health, or both. Often we simply push the costs of negative externalities forward to future generations and expect them to pay, not only financially but also in terms of their health and well-being. This failure to take responsibility for our broken relationship with nature has been going on for several generations, and the compounded costs of degradation are fast catching up with us. It's time to find a new way.

The costs associated with using nature should be distributed fairly across society and across generations. A fair distribution would mean that the polluter or user of an ecosystem good or service pays rather than the public at large. In the past, it was difficult to determine how an ecosystem service, such as clean drinking water from a mountain forest, can be maintained in perpetuity and how to price that service adequately. However, artificial intelligence enables us to measure and monitor biodiversity and its associated goods and services with increasing accuracy and allows us to determine who should bear the costs. AI is turbocharging our ability to allocate an exact price for an ecosystem service and to ensure that this cost is charged to the right user and reinvested in nature's ability to sustain this ecosystem service.

Running Dry

Prominent examples of a sustainable and equitable approach to structuring our economy include public water utilities, which have begun investing in ecosystem restoration to ensure water security.

For instance, the city of São Paulo in Brazil, with over 20 million people, faced a severe water crisis between 2013 and 2015. The city government joined forces with the private sector to invest systematically in the restoration and reforestation of critical freshwater ecosystems. The São Paulo Water Fund was established by a broad group of stakeholders and has conserved or restored 27,000 acres (11,000 ha) of the most important forest areas for the replenishment of local watersheds. The fund aims to further improve water security for the 26.7 million customers of the city's water utility.[16] Through the conservation and restoration of 165,000 acres (67,000 ha) of forest, the so-called Green Belt of Metropolitan Watersheds (*Cinturão Verde dos Mananciais Metropolitanos*) stabilizes ecosystem services at the landscape level and generates sustainable income to upstream communities.[17] When I visited this program in 2014, as a case study for the Global Partnership on Forest Landscape Restoration, it was already a clear win-win for farmers, water customers, and nature. Since that time, the city government and surrounding states have broadened the program. Funds for this essential investment are raised through water levies, city taxes, and corporate investments from companies that are highly dependent on water resources.

Honeyland

The economy of almond production in the US state of California is another example of where it would make sense to invest back in nature rather than ignore the costs of nature loss. The Golden State produces about 80 percent of the world's almonds. Almond trees are 100 percent dependent on bee pollination to bear fruit. California has around 1,600 natively occurring bee species, one of the highest diversities of bees in the world. There are only about 20,000 species of bees worldwide, making California the "Amazon basin of bee diversity," as former executive director of the California Native Plant

Society, Dan Gluesenkamp, puts it.[18] Despite this vast natural wealth of pollinators, native bees have become so rare that there are no longer enough of them to pollinate the almond trees.

The decline in almost all native bee species in California can be attributed to widespread use of pesticides, the lack of habitat, and the decline in diversity of flowering plants outside of almond season. Agriculture areas are simply too homogeneous and too poisonous for native bees to survive year-round in large numbers. Therefore, to sustain almond production, it is necessary to bring in about 2 million rental bee hives of European honeybees from across the United States, at a cost of almost 500 million USD.[19] European honeybees are one of humanity's oldest domesticated animals—they have been with us for over 9,000 years—and they know how to do their job. However, the rent-a-bee economy has its limitations and comes with costs and challenges, including the fact that European honeybees can further displace native bees. Also, commercial honeybee colonies themselves are starting to suffer from the widespread use of pesticides and the effects of climate change, resulting in the collapse and death of colonies on an accelerating scale. It would be much better if, for each almond and each fruit and vegetable produced in California, a small investment would go toward the conservation and restoration of native bee habitat.

More important, we should change our agricultural practices to ensure that nature can continue to provide enough clean water, fertile soil, and healthy bees to produce our food, including phasing out certain pesticides that contribute to bee decline. We will pay for the true cost of food and other commodities and products one way or another, and it is more transparent and fair and cheaper to pay when a problem first arises and is still manageable than when widespread ecosystem collapse is imminent or already underway. Honey, coffee, cocoa, and many other commodities have increased sharply in price in recent years; that is part of the cost of the decline of nature that we all pay. Would it not be much better if we invested the same amount into bringing nature back? By the way, the next time you eat honey, please think of the honeybees. One bee produces only about five drops of honey in her lifetime. Doing so is hard work, so let us pay nature back in the currency of respect, gratitude, and protection. Let's no longer take nature for granted.

The Indoor Generation

One of the most critical services of nature is related to something that most of us care deeply about: our health. A recent study in the science journal *Nature* showed that spending at least 120 minutes a week in nature is associated with good health and well-being. It also improves cognitive function and brain activity, lowers blood pressure, strengthens mental health, and enhances sleep quality.[20,21] The health benefits of nature are now so well documented that nature has officially been recognized as an effective approach to integrative healthcare in the United States, the European Union, the United Kingdom, Japan, and many other countries. Doctors in 35 US states are now prescribing contact with nature for a wide variety of ailments.[22]

And it is high time. A recent study across 14 countries found that the so-called Indoor Generation spends 90 percent of their time indoors, where the air can be up to five times more polluted than outside.[23] That is, unless you live in a city with particularly hazardous outside air quality. Living in damp and moldy homes increases our chance of asthma by 40 percent. We can all make a small shift in how much time we spend outdoors.

Now that we have read a few examples of the immense monetary value of nature to our economies and to ourselves, let's find out how we can trigger investments in nature at the level required to reverse nature loss.

The Oldest New Financial Asset Class

The mainstream finance industry is starting to anticipate the paradigm shift of assigning nature her true value in our economy, including the emergence of nature as a new financial asset class. Nature is of course not new as a financial asset class. To name just a few examples, every bit of timber used in the construction of a building, every bit of iron ore extracted from a mountain, and every fish caught in the ocean to feed humanity is nature's contribution to the economy. Nature is what underpins our entire economy; it's not a new and niche investment area. However, in the past, most financial asset classes related to nature were all about extraction: mining, fisheries, forestry, and even agriculture in its current predominant form is mainly a way of extracting nutrients and water. What is new is that nature is becoming an investible asset class where finance is specifically used to replenish, restore, and regenerate nature and to rebuild natural capital.

The finance industry refers to an asset class as an identifiable category of investment with similar financial characteristics, such as real estate. Forestry, fisheries, and agriculture have long been classified as "natural resource" investments. However, traditionally the gain of natural capital and the increase of biodiversity, or the sequestration of carbon, have not been included as metrics in any asset class. That is now changing, and nature—or, more precisely, the conservation, restoration, and sustainable use of nature—is now more clearly defined as a financial asset class, with better indicators of what constitutes "nature-positive investments." And that is important because pension funds and other large institutional investors are allowed to invest only in those assets that are properly classified and securitized, complete with a set of rules, tools, and metrics to determine risk and returns. All of those requirements are now being established for investments in nature.

More specifically, the finance industry is starting to pay close attention to nature as a "next big thing" in institutional investing. In September 2023, during the Opening Week of the UN General Assembly, I was sitting in the boardroom of one of the world's largest investment firms in New York City. A handful of other nature experts and I were seated around a huge round table of polished tropical hardwood, on the 78th floor of one of the Big Apple's tallest buildings. The firm had recently acquired a stake in a large sustainability consultancy, and it had developed an appetite for investing in nature. The company's leaders wanted to discuss how to best invest in this "new asset class" they had recently become aware of. They wanted to invest in nature-based solutions, in particular for climate change mitigation. I found myself in the company of powerful people who work at the heart of the world's financial industry, controlling billions of dollars in investments, and probably earning eight-figure salaries.

Looking out that day over New York Harbor, seeing the Statue of Liberty as a tiny dot in the distance far below, I wondered how I could explain to someone with this kind of view of the world the intricacies of, for instance, restoring a barren mudflat in Pakistan to a healthy mangrove ecosystem, or the plight of a subsistence farmer in Uganda, or why an Indigenous tribe in Indonesia would be trying to save the rainforests of Borneo from becoming palm oil plantations. So many of the rules of our modern economy and its markets are still stacked against sound nature investments that if we focused strictly on tallying up the financial costs and gains, more financial dividends would accrue by cutting down the rainforest, planting a palm oil monoculture, squeezing the soil dry for a few years, and moving on when the ecosystem collapses or catches fire. We need a few additional base rules for market economics to get investing in nature right. One such rule is to change how we account for natural capital not only on countries' balance sheets but on companies' balance sheets.

If You Can't Fight Them, Join Them

Since my meeting in New York, I have been in touch with dozens of new investment funds that are rushing to the aid of nature. The need for investor education is immense and urgent. Investors' understanding of nature as an asset class and the quality of their proposed projects vary widely. The entire investment community still requires significant capacity building on the subject of nature and, most important, on the social dimension of nature conservation and restoration. Fortunately, well-informed investors and campaigns have emerged from within the financial industry, underpinned by the necessary expertise on nature. In particular, the Task Force on Nature-related Financial Disclosure has brought ecological literacy and investment and reporting metrics for nature investments and impacts to over 500 companies and financial institutions with over 17 trillion USD in assets under management since it was established in 2020.[24]

With this growing recognition of the need to invest in nature comes a new challenge, though: There simply are not enough "bankable" projects of large enough size, in the 100 million USD and above range, to absorb significant capital. I believe this challenge will be solved in the coming few years, as the market matures and catalytic, smaller, and more risk-tolerant investments pave the way for more substantial sums by institutional, more conservative investors. This change will take some time, as successful conservation and restoration projects typically require a clear social contract between people and how they utilize common natural resources, such as groundwater or a community forest. These types of projects are often outside the usual expertise of bankers and investors. Nature as an asset class is different from most other asset classes because people are part of nature, which makes investments more complex. Compared to other investment areas, such as technology, nature requires more patience, considerable knowledge, and more moderate expectations on the financial rate of return. On the upside, nature investments done well are producing multiple benefits beyond financial returns, which makes them top contenders for so-called blended finance, or public–private partnerships, where some public or philanthropic funding is invested to catalyze private investments.

For billions and ultimately trillions of US dollars to be shifted into nature-positive investments, the mainstream finance industry must embrace this complex social dynamic and the ecological complexity of restoration. In a way, investing in nature is more about ego-system management than it is about ecosystem management. The social dynamics in a landscape are what makes or breaks a nature investment; the technical side of nature restoration is relatively straightforward. It is possible, though, and necessary to develop large-scale, socially and ecologically sound nature investments for commercial returns. Humanity has done more complicated things, such as building space programs; or the Burj Khalifa in Dubai, the world's highest building; or the internet with its thousands of data centers and millions of miles of undersea cables to connect all of us at lightning speed. Compared to some of the engineering and financing challenges we have already overcome, investing well in nature is relatively easy, once we put our best minds to it. Governments play a key role here because they can set the policy and market signals that indicate on which problems and solutions we should focus our creativity and energy. Governments also can fund the initial breakthroughs in innovation we need and help to overcome initial hurdles for private-sector investments into building and scaling an ecosystem restoration economy.

The good news is that banks and other investors are finally starting to move. There exists a fast-growing body of economic literature and ecological guidance on how to invest well in nature, at a holistic landscape level and for more than just financial return. My favorite investor framework for large-scale ecosystem restoration investments is the so-called 4Returns Framework by the Dutch Commonland Foundation, which encourages a minimum investment horizon of 20 years and aims for the additional returns of social capital, natural capital, and inspirational capital.[25] Large restoration projects and initiatives need funding, and a clear, compelling investment narrative will go a long way to mobilize the required capital.

A Billion Here, A Billion There

Just since 2021, private-sector nature finance has increased eleven-fold, to over 100 billion USD per year.[26] That is already about 10 percent of what is needed, and it is a good beginning.[27] With the right guidance and policies, we can ensure that this new investment interest in nature is sustained and builds a continuously growing industry based on clear rules and market signals. We can build nature investments into a trillion-dollar restoration industry, with a focus on holistic landscape conservation and restoration, as Willem Ferwerda, founder of the global restoration network Commonland, describes it.[28]

Fortunately, that industry is emerging, thanks in part to nature's unique ability to be a multi-solver: a solution that solves more than one of humanity's most pressing challenges, including the urgent need to mitigate and adapt to climate change. At the time of writing this book, at least 29 specialized, significant commercial investment funds focus on investing in nature conservation, restoration, or more sustainable land use (such as regenerative farming, which I describe more in Chapter 8). Most of these funds are less than a year old, and more funds are being established each month. Most have their origin in humanity's urgent need to invest in mitigating climate change as well as in the need for markets to play an effective role in linking supply and demand for carbon. Based on the physics of greenhouse gas stocks and flows, the carbon removal industry will become one of the most important industries of this decade and this century, so let us take a brief look at nature's potential to stabilize the climate and the market mechanisms that can be one tool to reach that potential.

Nature to the Climate Rescue

The climate crisis, which is inextricably linked to our relationship crisis with nature, has reached a stage of such urgency that an increasing number of decision makers in the economy feel compelled to act, even as concerted government action still lags. Basic physics tells us that reducing our emissions will no longer be enough to stay within a stable climate. As we saw in Chapter 2, even the current level of global warming we experience is neither safe nor stable. We will almost certainly overshoot the 2.7°F (1.5°C) target of the Paris Agreement on Climate Change soon and will then need to claw our way back into safe territory by removing carbon from the atmosphere. In addition to avoiding further emissions, we will need secure technologies and projects that suck carbon out of the atmosphere and store it safely in plants, the soil, rocks, or deep underground.

The ability of the private sector to quickly innovate and scale climate and nature solutions was one of the reasons I joined global technology leader Salesforce in May 2022 after a long career at the United Nations. Salesforce has been active in identifying and scaling climate solutions for over a decade, including through philanthropic grants, venture capital investments, and its purchasing power. Our Impact Team at Salesforce is, for example, investing in so-called carbon dioxide removal technologies, such as enhanced rock weathering, direct air capture, or carbon capture and storage to draw down CO_2 from the atmosphere and store it safely. Although developing and scaling such technologies are essential, no human-made technology is yet even remotely as sophisticated, cheap, beneficial, and scalable as nature. This is why Salesforce is also investing in nature-based solutions for climate change. Intact forests ecosystems, in particular, are essential for stabilizing the climate, both in terms of halting deforestation and restoring forest landscapes at a significant scale. And reforestation and forest conservation projects, if done well, produce many social and ecological benefits beyond climate mitigation.

Trees and other plants, including phytoplankton, have been absorbing carbon from the atmosphere for billions of years and have become very efficient at it. Nature-based solutions for climate change should be a major focus of our investments to mitigate the climate crisis. Nature has the potential to provide a large part of the solution to draw down more carbon, and it also provides numerous other benefits to humanity. The other necessary focus of climate action is, of course, a swift and determined decarbonization of our entire economy and transitioning all energy production to renewable sources. If we do just those two things, we will have solved not only the climate crisis but issues of equality, energy security, energy access, clean air, human health, water security, and many others. Nature is indeed a multisolver. So what do these nature-based climate projects look like, and who is already investing in them?

The Carbon Markets

On the demand side, a growing number of companies and countries are interested in investing in carbon projects that can both mitigate climate change and provide additional benefits, such as improved water security, biodiversity conservation, and enhanced climate resilience. They look to purchase emission avoidance or reduction units, known as carbon credits, which represent 1 ton of carbon that is prevented from entering the atmosphere or is removed from it. As carbon markets are still young and maturing, the price of carbon credits varies widely, as does their quality. In recent years, the markets have become increasingly sophisticated in terms of transparency, ease of access, and the credibility of the products. Carbon markets play a crucial role in matching supply and demand in the most cost-effective manner. We differentiate between so-called compliance markets, like the European Emissions Trading Scheme, in which companies are forced to participate by law, and the voluntary carbon market, which, as the name implies, has corporate buyers who invest in carbon credits for a variety of voluntary reasons, mainly as an additional opportunity to lower their carbon footprint. Individuals can also purchase carbon credits from several platforms, including from the United Nations.[29]

The relatively new voluntary carbon market has already experienced several boom-and-bust cycles in its short lifespan. This has resulted in rapid learning and market improvements. We need this market to function effectively and scale rapidly as a tool to channel finance to the most cost-effective solutions. We also need voluntary investments in carbon markets urgently right now to test and develop new technologies and new solutions to effectively channel the larger amounts of funding that inevitably will be made available to solve the climate crisis. It is likely that more regulated markets and so-called compliance markets will emerge, with larger financial flows than currently in the rather small voluntary carbon market. Key lessons need to be learned fast and sometimes the hard way. In 2023, an investigative piece in *The Guardian* and other major newspapers claimed that a lot of carbon credits were not credible and basically worthless.[30] The methodology of their investigation has been contested.[31] Still, the publication led to a crisis of trust in voluntary carbon markets and to a frenzy of activity to improve them, including the development of better rules and technologies for transparency, the establishment of independent rating agencies to assess the quality of projects, and a clearer regulatory framework. In the long run, a strongly regulated market with mandatory participation from specific industries and countries is the most likely future for carbon credits.

Carbon markets are complex and require clear rules, continuous monitoring, and ongoing improvement, like most other markets. Despite the recent dip in credibility, they are projected to reach more than 7 billion USD per year by 2030 and between 45 and 250 billion USD by 2050.[32] The growth is driven in part by new nature targets, which can be achieved in a "two for the price of one" purchase of high-quality carbon credits from nature-based solutions, such as reforestation or agroforestry. Demand for such high-quality carbon credits is strong and receives clear market signals.

In 2024, Google, Meta, Microsoft, and Salesforce announced the establishment of the Symbiosis Coalition to purchase at least 20 million tons of carbon credits from high-quality nature restoration projects, such as forest restoration and agroforestry, at a market value of at least 200 million USD. There has been considerable concern among civil society about the motives of large companies to invest in carbon credits, with some viewing it as a get-out-of-jail-free card, meaning that if companies invest in carbon credits, they can simply sit back and do little else for climate action. However, recent research demonstrates that the companies investing in carbon credits are decarbonizing their entire operations and value chains on average twice as fast as other companies.[33] In other words, it is not an either/or: Buying carbon credits and reducing emissions simultaneously as quickly as possible is a sign of ambitious climate leadership. More companies can and should do both, and carbon credits should be included in any credible corporate climate action plan as a last, but not later investment.

The Ecopreneurs

On the supply side, an entire army of startup companies, investors, project developers, and nongovernmental organizations is generating new projects. Over the past years, these "ecopreneurs" have developed almost 10,000 carbon projects and registered them across the 12 largest international crediting registries.[34] Many of these projects invest in the conservation and restoration of nature. Because when it comes to absorbing and storing carbon at a large scale and low price, nature is hard to beat. After all, she has been implementing and improving carbon storage technology for 3.5 billion years. The world's forests alone contain about twice as much carbon as the entire atmosphere, and the world's soils contain about three times as much. The ocean has absorbed approximately 25 percent of all human CO_2 emissions to date, and, even more important, it is absorbing a significant portion of the excess heat generated by greenhouse gases.

The ocean has absorbed more than 90 percent of the excess heat in the climate system since 1970. This borrowed time from the ocean may come to its limits soon, though: There are indications that oceanic heat absorption has reached capacity, and higher CO_2 levels have already made the water more acidic, with detrimental effects on marine life.[35] Similarly, terrestrial carbon sinks like forests are also starting to show signs of climate-related stress, which could turn them from carbon sinks to carbon sources, where forest fires and drought cause more greenhouse gases to leak from the system than are being absorbed.[36] Nature is a key ally in our race to slow climate change, yet it is also increasingly vulnerable to the impacts of a changing climate.

This sense of urgency is exactly why we need carbon markets as one tool in our climate action toolbox. Given the right market signals and policy conditions, market mechanisms can operate and grow remarkably quickly. Just recall the global COVID-19 pandemic and how the shortage of face masks and vaccines was overcome in record time. Currently we are facing a similar health emergency for Planet Earth. As that becomes clearer to more and more of us, the speed of the response is also growing. Carbon markets can play a critical role in closing the estimated 1 trillion USD annual financing gap for nature. By channeling private-sector funds into nature-based solutions, the market provides additional resources beyond what national governments can offer, thereby enhancing the scale and effectiveness of conservation efforts.

In addition to carbon credits, the Global Biodiversity Framework adopted in December 2022 foresees the development of so-called biodiversity credits. However, that market has yet to prove that it can reach a significant scale. Combining nature and climate objectives into high-quality carbon credits might be a more promising approach to achieving the speed and scale of nature investments than developing separate markets for carbon and biodiversity credits. Carbon or biodiversity markets alone will not be enough, though. Even if carbon markets reach 35 billion USD by 2030, the current most optimistic projection, and biodiversity credits perhaps a fraction of that, it will still be far from enough investment in nature. We need a step change on an entirely new level of investing, and that means combining public and private funds in innovative ways.

Nature as Critical Infrastructure

We will reach the required investments in nature, approximately 1 trillion USD per year, once the public sector starts to view nature as critical infrastructure for our societies and invests accordingly. Just like roads, airports, and power lines, the many ecosystems of nature provide essential services to our societies. However, currently we take the services from this natural infrastructure for granted.

There are now more and more areas of our life where nature has been lost to such a degree that her absence starts to be noticed by others than nature lovers, including in places where nature provides essential infrastructure and public services for human civilization, such as clean drinking water, flood protection, clean air, storage or carbon, and healthy and fertile soils.

Let's take the example of a more well-known infrastructure category to illustrate what this step change of viewing nature as critical infrastructure for humanity would mean. German highways are famous for their lack of a speed limit.[37] They are solidly planned and constructed, and that comes at a high price: 1 kilometer of highway (0.62 miles) costs between 6 and 20 million USD. And Germany has 13,200 kilometers (8,200 miles) of Autobahn, so establishing this road network has cost somewhere between 79 and 264 billion USD, not counting annual maintenance and repair costs. And this is just one part of the infrastructure costs of running a modern society in a single country. Power grids, data centers and networks, water utilities, railroads, and airports are all accepted as necessary investments. What if we began to view nature as critical infrastructure that requires these levels of investment too?

Let us stay with Germany as an example before we explore a few other countries. Germany, like all countries that signed the Paris Agreement on Climate Change, has committed to reducing its emissions and is almost on track to do so, largely due to a rapid shift to renewable energy. However, Germany needs to do more to achieve its target of a 65 percent reduction in greenhouse gas emissions by 2030 and an 88 percent reduction by 2040. A surprising 7 percent of the country's emissions come from peatlands (an ecosystem you might also know as moor, fen, marsh, or mire). Peatlands have extremely high carbon content, and, for centuries, they have been drained and plowed, and peat has been extracted for fuel or as a medium for soil improvement in agriculture in most countries across the Northern Hemisphere. To reach its climate goals, Germany will have to reduce the 7 percent of its national emissions from peatlands to near zero. That is exactly the aim of its National Peatland Strategy, which aims to restore 4.5 million acres (1.8 million ha) by rewetting them, basically turning them back into wetlands and reversing centuries of drainage. In economic terms, that is a bargain for Germany's climate actions, as replacing the agriculture currently taking place in those areas, which has low economic value, with restoration activities can cost around 4,000 USD per acre. You could think of the peatlands and other critical ecosystems, as Germany's climate stabilization infrastructure. Compared to overall public spending, the cost would be manageable. For nature, it would mean an unprecedented influx of funding. The investment of upward of 18 billion USD could be a combination of a public works project with private finance, including for carbon credits, and income from tourism.

A Restoration Boom

Public investment in nature could trigger a restoration boom and unlock private investments in vibrant new industries in areas that are often remote and in need of infrastructure and development funds—a win-win for the climate and the rural economy. At the same time, restoring peatlands would secure an essential water supply for future generations. Currently, Germany does not yet have the vision or public support to invest in nature at this level of spending, which is typically reserved for infrastructure projects or the military. However, I believe that with growing pressure to react to the climate and nature crisis and with better economic business cases for restoration investments, these types of large-scale projects will emerge around the world, whether it is restoring the productivity of the Atlantic Rainforest in Brazil; securing mountain forest watersheds in central Sri Lanka to maintain national tea production; rebuilding the coral reefs of the Great Barrier Reef in Australia, as one of the country's top tourism destinations; or restoring the Everglades in the United States and mangrove forests around the tropics as key coastal infrastructure, protecting many trillions of dollars in coastal real estate. We are on a good and necessary path to see nature restoration grow into a trillion dollar industry.

Big, Bold Bets for Nature Finance

Before we can grow nature restoration into a trillion USD industry, we need to recognize that many of the world's major ecosystem restoration opportunities have an investment gap, especially in the early years of restoration. Effective ecosystem restoration techniques, such as agroforestry, regenerative farming, or changing land use to generate carbon credits, often incur high up-front costs, and the techniques start to generate financial returns only after several years— often 5 years or more in the tropics and up to 10 years or more in temperate climates.

Trees in agroforestry investments, for example, typically require several years before they produce nuts, fruit, or timber. During this period, continuous investments in pruning, planting, and management are needed, and little or no cash flow is generated. If we add to this cost to the often high-risk profile of nature-based solutions, such as the risks of storms, fires, or social or political changes across a landscape, the investment case becomes challenging. Once a critical mass of investments has been made to mitigate or better understand and price these risks, easier investment flows and better market returns will be generated. It is relatively easy to raise initial small sums of money to get a nature restoration project started, often from philanthropy, but it is much more difficult to raise low-interest, long-term capital for large-scale projects. The asset class of nature is still so new and not well understood, and due to its high perceived and some real risk, investors expect double-digit percentage points in return on their investments. This situation makes the required capital for many large-scale restoration investments prohibitively expensive.

Smart, risk-tolerant philanthropic capital can have high impact and leverage by helping to derisk and catalyze larger commercial investments. Grant funding should be available early and easily for project development and then again to derisk the projects as they progress toward commercial viability, which can unlock significant commercial investments. Grant capital can be provided as so-called reimbursable grants, which are low-cost loans with a high risk tolerance. If the project is successful, the grant is repaid with low or no interest, and the funds can be reinvested to support the next promising project. If a project is not successful, the grant is not paid back. Still, in that case, valuable lessons have been learned about project approaches and a foundation may have been laid in a given landscape or seascape for a more economically viable future approach. In many areas, early experimentation and learning are necessary before large institutional investors, such as pension funds, will write investment checks for nature restoration in the tens or hundreds of millions of US dollars.

Brazil's Restoration Boom

In some countries, and in sectors such as large-scale reforestation, projects are already relatively mature, and nature as an asset class can attract significant institutional investments. The investment strategy of BGT Pactual Timberland Investment Group, to give just one example, is successfully mobilizing 1 billion USD to restore approximately 329,000 acres (133,000 ha) of natural forest and establishing sustainable commercial tree farms on an additional 329,000 acres in Latin America, including Brazil's highly deforested Cerrado region. The combination of commercial timber plantations, a well-understood asset, with ecosystem restoration for carbon and biodiversity credits, as a new asset class, is compelling enough for institutional investors to provide capital at reasonable rates of return.[38]

Even where nature investments are becoming more mainstream and less costly in terms of borrowing capital, still often lacking are the so-called enabling conditions, such as the right policies, laws, and regulations to create long-term investment security. The most catalytic philanthropic investments in large-scale nature restoration include targeted policy advocacy for improved conditions for nature investment, such as clear carbon markets. The Salesforce Nature and Sustainability Fund, established in 2021, has helped to derisk private finance and to create solid enabling conditions for nature investments in countries such as Brazil. Salesforce's grantees have supported, for instance, clearer policies about mangroves in Brazil, and the company has supported, with many other companies and industry associations, the establishment of a carbon law in that country that will unlock new domestic and foreign direct investment.

Another area where philanthropy for nature is essential is the building of ambitious, action-oriented coalitions. The Global Mangrove Breakthrough is a collective effort to mobilize 4 billion USD in private and public funding to conserve and restore 37.5 million acres (15 million ha) of mangrove forests around the world.[39] A few years ago, it was just an idea, but thanks to a grant from Salesforce, UBS, and other donors, it is now supported by hundreds of nongovernmental organizations, governments, and corporations and has a clear financial roadmap to increase investments in mangroves worldwide. Investments of 4 billion USD for mangroves are possible if governments start to view them as critical infrastructure for coastal protection and if more governments and corporations invest in predictable, transparent carbon markets. These changes can be significantly accelerated through advocacy, scientific research, and pilot projects funded by grants and soft loans.

These types of catalytic philanthropy need courage, because some of them can fail. We need the courage to be fast and unbureaucratic, and aim for systemic change, which means being willing to lose some money in testing approaches that might not work and sharing the lessons with the world at large while also sharing the commercial success of approaches that work with the world at large. Philanthropic funding invested as concessional capital—capital made available at below-market prices and conditions—can reliably leverage 4 USD for every 1 USD invested and in some cases more than 10 USD for each dollar invested. Thus, there are good business cases for deploying philanthropy to unlock the needed much larger public and private funding flows of up to 1 trillion USD per year toward nature.[40]

All the Money in the World

As we have seen in this chapter, there is a clear business case and an emerging asset class for investing in nature—for governments, for companies, and for society at large. The financial business case becomes clearer and better articulated the more we understand nature as an asset class and the more we level the playing field by shifting harmful subsidies. However, having a strong business case alone is not sufficient to move toward planetary-scale nature conservation and restoration. I have often been asked about nature's "business case" by individual businesses and at local to international political levels. There usually is a good business case to make if we take the true value of nature into account.

However, deep inside, I am somewhat bemused by questions about the business case for nature. To me, it is like someone saying "So, I have heard that being a good husband and a good father is recommended. What is the business case?" Or imagine someone enthusiastically explaining to you that they have just calculated the monetary value of their children and their spouse, and they found out that they are worth a lot of money to them personally and to society at large. Would that knowledge make you take better care of your family? Indeed, if you neglected them and let them starve, it would be a personal and macroeconomic financial loss. Does this sound strange? Well, that is how nature finance studies sometimes sound to me, and probably to the majority of the world's Indigenous peoples, who balk at the idea of reducing nature to a monetary price tag.

Recent studies by the World Economic Forum demonstrate that more than half of the world economy is dependent on nature.[41] Although I am grateful to the authors for crunching these numbers and the report is useful in some ways, the findings are also absurd. Without nature, life on Planet Earth simply would not be worth living. What is the point of calculating exactly how much economic activity depends on nature? The entire world economy is a 100 percent subsidiary of Mother Nature. We eat nature's produce for breakfast, lunch, and dinner. Of course we all depend on nature, all the time. That means our economy, which should be a tool in service of society, is also fully dependent on nature. Giving nature a monetary value is useful only to a certain degree. More important is that we give nature more respect and gratitude.

Let's Do the Right Thing

Certain reasons to invest in nature are more fundamental than an economic business case, just as being a good person is about more than economics. Being a good spouse might save you an expensive divorce and significantly increase your happiness, which in turn will lower your healthcare and insurance costs. Being a good mother or father means your children are more likely to care for you when you are old, and investments in their education will mean they have a higher income and can later in their lives pay more taxes to keep the necessary infrastructure going that sustains you in your old age, such as hospitals and public transport. These are all good reasons why you should be a good spouse or life partner as well as a good parent. However, there is a much more profound and important reason why you should strive to be a good person. Our relationships define us. They make life worth living and give us purpose and meaning.

Our bonds with others define what it means to be human. Our loved ones and other close relationships inspire us to become better people. It is like that with nature too. In my relationships, I include not only my family, friends, and colleagues; I am also building a relationship with the land I live on, with the teeming life in our garden, and with the wildlife in the forest surrounding our farm. On a rational level, I understand that we need healthy soil and high biodiversity to ensure a clean drinking water supply. Yet, on another level, the smell of dark, rich humus is simply something that makes me happy and fulfilled. And that is priceless, and a much better incentive to act than a monetary transaction. We have to be careful not to overstretch the economic argument for investing in nature, lest we replace or dilute a more powerful motivation: that we want to build a healthy relationship with nature and with each other.

Beware of Your Motive

Numerous studies have shown that shifting interpersonal transactions from a social contract to an economic level can reduce the sense of responsibility people feel in particular situations. In one study published in 2000, Uri Gneezy and Aldo Rustichini conducted an experiment at a group of daycare centers in Haifa, Israel, and found that introducing a monetary fine for late pickups increased the number of parents arriving late.[42] Before the fine was implemented, parents felt guilty about being late. However, once the monetary penalty was introduced, parents could "pay" for their tardiness, effectively absolving them of their guilt. In other words, this shift from a social transaction, governed by moral and social norms, to a market transaction made it easier for parents to justify their late arrivals.

This phenomenon aligns with the concept of the crowding out effect, where extrinsic motivators, such as fines, can undermine intrinsic motivation, including the desire to be considerate and punctual. I realize that this hierarchy of motivations applies more to individual human beings than to other key actors, such as companies, banks, or local and national governments, which are expected to act on a clear and rational business case. But even for them, the power of a new social contract and relationship between humanity and Earth should not be underestimated. We need to factor a new relationship with nature into our business cases, and a deep relationship with nature will serve as a powerful motivator for individual action that can drive social change more effectively than simply putting a price on nature and increasing its market value could.

According to Oscar Wilde, "A cynic is a man who knows the price of everything and the value of nothing." I realize that in our utilitarian, secular world, money is the language that is most widely spoken and understood by decision makers. And it is important to know that the economic arguments for nature are compelling. However, we can have a powerful and exciting relationship with nature beyond financial gain or loss. In the next chapter, join me to explore explore people and places around the world where this kind of relationship still exists or is being rediscovered.

Chapter 5

The World Plus 10 Percent

The Earth does not expect you to save her, she expects you to respect her. And we, as Indigenous peoples, expect the same.

—Nemonte Nenquimo[1]

Western and Indigenous worldviews have clashed for centuries. It is time to reconcile them. Life as we know it is at stake on our planet due to a compounded crisis of climate change, nature loss, and pollution, and we all can feel the impacts. Pollution is causing a rise in pulmonary and cardiovascular diseases and certain forms of cancer.[2] The rise in the use of plastic bottles has been linked, for example, to the early onset of colorectal cancer.[3] Studies have found microplastics in human blood, lungs, liver, kidneys, breast milk, and the brain, with concentrations in the brain being particularly high. Recent studies indicate we could each have up to a quarter of an ounce (7 grams) of microplastics in our brains, about the weight of a plastic spoon.[4,5]

At the same time, climate change is accelerating and exacerbating the occurrence of billion-dollar natural disasters, driving up the cost of insurance for real estate and other essential economic sectors. In the United States, natural disasters causing damage of over 1 billion USD have increased sharply in recent decades, from around three events on average per year in the 1980s to over 20 per year in the early 2020s, with massive costs to human lives, health, and property.[6] And our natural world is imploding, with a million species at risk of extinction. Something's got to give.

With the right shift of perspective and paradigms, we can learn to live within our planetary boundaries. Doing so will enable us all to lead better, healthier, and more fulfilling lives. It is possible, and the wisdom and knowledge of Indigenous peoples and local communities can teach us much about our urgent task of making peace with nature and building a society with an economy that allows us and generations to come to live lives worth living on a finite planet. However, in addition to basic ecological knowledge, a higher level of trust in life and in the processes that underpin nature is required. In this chapter, we cover what we can learn from Indigenous peoples and local communities about a healthy relationship with nature.

Take No More Than Half

A guiding principle in North American Native ecological knowledge is never to take the first or the last of anything in nature and generally never to take more than half of any one animal or plant in a hunt or harvest. In her essay "The Honorable Harvest," Robin Wall Kimmerer, a biologist and citizen of the Potawatomi Nation who grew up in upstate New York, illustrates this principle through the story of Ojibwe wild rice harvesters, who leave a significant portion of the rice in the water to allow it to reseed and provide for wildlife, such as ducks.[7] What is left behind is not wasted but sustains the ecosystem for future years. This guideline is part of a broader set of Indigenous ethics known as the Honorable Harvest, which also includes asking permission, taking only what you need, and leaving some for others.

No such restrictions are inherent in the logic of unfettered capitalism, unless they are introduced through regulation. The goal of our global economy seems to be endless growth and individual profits—the more and faster, the better—with no limits in sight, regardless of the costs to society at large or to future generations. The motto of capitalism in its current form could be "I Want the World Plus 10 Percent," which is the opposite of the Honorable Harvest. Because of the extractive mindset that takes without giving back, we have come surprisingly close to the predictions of the first-ever large-scale environmental forecast: The 1972 *Limits to Growth* report commissioned by the Club of Rome used computer modeling to explore the consequences of continued economic and population growth on a finite planet.[8] As predicted over 50 years ago, we are now reaching the limits of conventional economic growth.

The planetary crisis we find ourselves in does not surprise many of the world's Indigenous peoples. The Indigenous worldview offers a profound lens through which to understand life, one that is rooted in interconnectedness, spirituality, and respect for the natural world. Unlike the compartmentalized perspectives often found in Western thought, Indigenous ways, as well as many Eastern philosophies such as Taoism, emphasize living in harmony with the natural world and embracing life's natural flow. At the core of what is sometimes called the cosmovision of many Indigenous peoples is the belief that all living beings—humans, animals, plants, and even the land itself—are deeply interconnected. This interconnectedness is not merely metaphorical; it is a lived reality that shapes every aspect of Indigenous life. The spirituality and mythology of Indigenous cultures reflect these principles of interconnectedness and reciprocity.

All human mythology contains cautionary tales about the over-exploitation of nature. Yet, in our modern and presumably "enlightened" age of science and technology, we seem to have forgotten these tales. We are like the farmer in the tale of the goose that lays golden eggs, a fable originating in ancient Greece.[9] A poor farmer owns a goose that suddenly begins laying golden eggs. At first, the farmer is overjoyed and becomes wealthy from selling the eggs. However, as his greed grows, he becomes impatient with the goose's slow egg-laying rate. Believing that the goose must contain a large quantity of gold, he kills it, hoping to get rich quickly. Instead, he finds that the goose is just like any other, and he loses his source of golden eggs, ultimately becoming poorer than before.

Similar stories exist throughout human history and the world. For example, the tale of the Drought Bird from the Kalahari tells of a bird that consumes all the water in a river, ultimately leading to the river's disappearance and the surrounding land becoming barren. In more modern times, Dr. Seuss's *Lorax* is a fable that illustrates the consequences of uncontrolled growth and reckless exploitation of natural resources.[10] It depicts the destruction of an ecosystem due to unchecked industrial expansion and the near extinction of species. We have become impatient with nature's slow rate of producing the goods, capital, and services we want. We want all the golden eggs, and we want them now. Nature is buckling under our mindset of exploitation. But there is another way.

Convergence

In an Indigenous worldview, nature is alive, vibrant, and deserving of care and reverence. Rivers are ancestors, mountains are protectors, and animals are kin. This perspective fosters a sense of mutual respect and reciprocity between humans and nature. Harmony with Earth is not just an ideal; it is a lived responsibility. Although that might sound like a romanticized view of nature in our secular and hectic modern world, the strong connection to the place we call home is widespread in many communities and cultures. It is not limited to Indigenous peoples and their lands and territories.

Equally important is the emphasis on community and relationships. Knowledge in Western societies tends to be compartmentalized into specialized fields and grounded in empirical evidence or singular truths. In contrast, Indigenous knowledge systems encompass multiple truths, shaped by lived experiences and oral traditions. Indigenous systems are fluid and adaptive rather than rigidly fixed. Western and Indigenous worldviews differ not only in their ways of understanding life but in their ways of living it. Indigenous worldviews call for stewardship—a relationship with Earth based on care rather than control. Relationships—with family, with community members, and with the natural world—are central to existence. Individual identity is deeply interlinked with the collective identity. Harmony within these relationships is nurtured through practices of reciprocity and mutual care. The well-being of one is tied to the well-being of all.

I do not advocate that our entire Western civilization and our modern market economy adopt an Indigenous worldview. Doing so would be difficult, given how far removed we are from a deep reciprocal relationship with nature. However, I do believe that those of us who consider ourselves heavily influenced by Western thought and culture can learn much from the Indigenous worldview for building a better, more resilient, more connected, and more wholesome society, with an economy that serves our needs rather than us serving the needs of the economy.

In the next section, we explore some areas where human relationships with nature are the strongest on Earth and based on mutual respect. Where do human culture and nature coexist, fusing into a new and mutually beneficial whole? And what happens when that relationship unravels? Is it nature that suffers most, or us? We discuss some examples where the Indigenous and Western worldviews converge toward a harmonious relationship between nature and people, grounded in traditional ecological knowledge and modern science. Let's start in one of Earth's most breathtakingly beautiful places, on the mountaintops of South America.

A Movement Is Born: Acción Andina

In July 2011, I was hiking up the slopes of Cotopaxi, Ecuador's highest mountain. It took me almost half an hour to climb the few hundred steps from the road access point to reach the Refugio José Rivas, a historic mountain shelter that was the destination of our small group of government officials and UN staff. At an altitude of 15,091 feet (4,600 m), the air is thin, and breathing becomes hard work. Every step is an effort. It is here, across the high Andes of South America, that an extraordinary restoration success story is unfolding. Over 35,000 volunteers—a number that is growing rapidly—work year-round on the steep mountain slopes of Argentina, Bolivia, Colombia, Chile, Ecuador, and Peru, with plans to expand to Venezuela. During the wet season, which lasts from November to March, thousands of people plant native trees. Since 2018, they have planted 12 million trees in more than 500 sites across six countries. And this is only the beginning. The social and ecological restoration movement Acción Andina is bringing together the remote communities across the national boundaries of the Andean countries and taking the highest parts of South America by storm. After only a few years, the movement is changing the face of the continent.

A Magical Tree

The high-altitude Andean forests are some of South America's most critical ecosystems. In the cool, moist mountain climate, these forests collect, store, and release clean water for millions of downstream people, animals, and plants. As the other primary drinking water source of much of the continent, the Andean glaciers, shrinks due to climate change, the forests have become even more important as a source of water. Growing usually above 11,400 feet (3,500 m) and up to 16,400 feet (5,000 m), the unique woodlands stretch from Venezuela in the north to Patagonia in the south of the continent.[11] The forests are often shrouded in clouds and consist primarily of trees of the genus *Polylepis*, which is comprised of about 30 known species of trees and shrubs. *Polylepis* trees are known by various names in different local languages. They are slow-growing, gnarled, and often covered in lichen and mosses. The forests they form look like something out of *The Lord of the Rings*: a magical ecosystem of immense age and significance.

Today, only between 1 and 10 percent of the original forest cover remains, depending on the country and location. Water for communities downstream is becoming scarce. For example, Lima, Peru, is one of the driest megacities in the world. Acción Andina wants to reverse the loss of forests. The initiative has a bold vision of restoring 2.47 million acres (1 million ha) of *Polylepis* forests across the Andes over the next 25 years—an area approximately the size of the island of Jamaica, or about 1.4 million football fields. The many communities behind the initiative aim to plant over a billion trees. And they are well on their way. What drives these communities, often living under harsh conditions, to spend so much time and effort on ecosystem restoration?

The Roots of Acción Andina

The movement started with the visionary idea of Indigenous leader Constantino "Tino" Chutas Aucca, who was born in Cusco, Peru, in 1963. He is a direct descendant of one of the families that founded the Inca Empire. His second name, "Aucca," means "warrior" in his ancestral language. And he certainly lives up to his name. Tino has been a tireless advocate for climate justice and the restoration of nature and water benefits among South America's most vulnerable communities across the high Andes for over 40 years.

I first met Tino in another iconic mountain range, in the Swiss Alps, at the 2022 annual meeting of the World Economic Forum in Davos. This small mountain town has hosted one of the most exclusive gatherings of world leaders and corporate executives for over 50 years. Tino's vision to restore all the high Andean forests was being announced as a flagship program of the global Trillion Trees initiative (1t.org), which was founded in 2020. We had invited Tino and his fellow Acción Andina leader, Florent Kaiser, to a celebration of the initiative's success two years after its launch. Tino, now in his early 1960s, exudes the confidence of someone who has found his true purpose in life. I was immediately impressed with Tino's and Florent's infectious enthusiasm and their single-minded focus on generating support for the vulnerable and often impoverished communities in the high Andes.

Florent, a French–German national who emigrated to South America, is the chief executive of an international nongovernmental organization (NGO) called Global Forest Generation. Tino inspires and leads the people and communities on the ground, and Florent and his team drive the strategic planning and organize international funding and support for Acción Andina. In a strong intergenerational and north–south partnership, Tino and Florent are brothers in spirit in this exciting adventure (see Figure 5.1). Since our event at Davos in 2022, Tino, Florent, and their respective teams have received multiple environmental prizes and awards, representing both their personal and their organizational success in developing Acción Andina. Awards include Prince William's Earthshot Prize, the UN's World Restoration Flagship award, and Tino's Rolex Entrepreneur Award.

Figure 5.1 Constantino Aucca Chutas (president of Acción Andina and ECOAN) and Florent Kaiser (colead of Acción Andina and CEO of Global Forest Generation) at a reforestation event in Ecuador in October 2023.
Credit: Photo by Todd Brown/UNEP

This stellar rise to international recognition is rooted in the work of the Peruvian conservation nonprofit Asociacion Ecosistemas Andinos (ECOAN), co-founded by Tino, which developed a sophisticated reforestation model built on more than 20 years of experience with local communities. Early in his career, Tino recognized the power of applying to ecosystem restoration the cultural principles of reciprocity and community service that had helped to build the Inca Empire and have kept local communities alive for thousands of years in the thin air and inhospitable climate of the high Andes.

A Culture of Reciprocity

It seems paradoxical that local communities with very limited eco-nomic resources would participate so enthusiastically in the hard work of reforestation. The reason lies in a deep connection to the land and a practical understanding of the importance of healthy watersheds. Many of the communities involved are Quechua-speaking descend-ants of the Inca. They value and practice the ancient tradition of *Ayni*, the sacred reciprocity between people and nature, and *Minka*, collective shared community service for the common good. This con-cept of reciprocity with nature and with everyone in the community is the key to understanding the success of Acción Andina. The move-ment is not just about planting trees. It is about working together in a spirit of reciprocity to ensure the well-being of the whole community and the land that sustains it.

"It's a Party!"

Hundreds of community members dressed in their colorful traditional garb danced, sang, and played drums and other instruments on their way to the annual Acción Andina tree planting festival in Cusco, Peru. At these festivals, every community member first engages in traditional singing and dancing, a rare opportunity to revive their cul-tural heritage and celebrate Pachamama, Mother Earth. The festivities energize people of all ages for the work ahead; it is part of the moti-vation needed for shouldering thousands of *Polylepis* tree seedlings, which were nurtured for over a year in community-run tree nurseries, and carrying them on a long uphill climb. Because the soil at the planting sites is usually too compacted for natural regrowth and remnant forest patches with tree seeds are often too far away, the young trees have to be planted by hand. Once at designated res-toration sites, the people dig holes in the dense soil to plant the trees.

This is hard work even at sea level, but in these high altitudes, I could not match the pace of the women, men, and even children who planted trees for a whole day.

Drone footage of the first planting season, with thousands of volunteers of all ages planting trees as far as the eye could see, on very steep slopes, that Florent and Tino shared took my breath away. At one site, volunteers planted 50,000 trees in a single day. Since that time, several communities working together have planted 100,000 trees in a single day. In the 2023–2024 season alone, communities planted over 3.4 million native trees across 2,923 acres (1,183 ha), and they have planted over 120 million trees altogether since the movement began in 2018. "It's a party. We're planting happiness!" says Tino, describing the spirit of celebration of culture, of community, and of reciprocity with nature.

More Than Trees

The holistic forest ecosystem restoration, conservation, and community development strategy of Global Forest Generation guides Acción Andina's approach. Activities include the scientific mapping of remaining forests and the species that depend on them to prioritize the protection of remnant *Polylepis* forest patches, mainly through conservation agreements with local communities. Acción Andina supports community investments and social services, including dentist and doctor visits, access to medicine, and solar panels. The program thereby not only restores the ecosystem but also revives the strength of the community, enhances the viability of Indigenous cultures, and improves people's quality of life. To ensure the long-term success of Acción Andina, Global Forest Generation provides capacity building and support for local leaders, project coordinators, and nursery staff.[12]

The funding for Acción Andina comes from various private donors and corporations through philanthropic grants. Because this type of funding typically is volatile, as corporate priorities can change quickly, Global Forest Generation is developing a substantial and long-term trust fund for Acción Andina to receive donations from both public and private sectors to establish a more stable and predictable funding source for the movement's growth. Given the significant positive impacts already achieved, this investment would set up the entire region for a restoration not only of essential ecosystems but also of culture, social cohesion, and economy.

Acción Andina also provides economic support to the participating communities through technical and financial assistance for sustainable agriculture, ecotourism, and microbusinesses while helping communities secure land titles and protecting them from exploitation. The high Andes are of interest to international metal and mining companies. Copper, gold, tin, iron, rare metals, and the new gold of our time, lithium, are all found there. Although mining operations have come under more public and regulatory scrutiny recently, they still often leave local communities without clear benefits. Instead, local communities often carry the burden of environmental degradation and resource competition. Lithium extraction, for instance, consumes vast amounts of water, lowering water tables and drying up rivers, lakes, and wetlands. The support that Acción Andina provides to local communities, such as access to alternative livelihoods, international attention, and access to a network of like-minded people across the Andes, also enables the communities to be more confident and better informed in their dealings with powerful external actors and pressures. The restoration of watersheds, in turn, makes it more feasible for mining companies to have long-term access to water without undue competition for resources with local communities and agriculture.

The Scale of an Empire

A strong and growing local conservation and restoration partner network is key to Acción Andina's success. During the 2023–2024 season, 15 regional partner organizations across 22 project landscapes collaborated with 117 local communities, reaching an estimated 35,000 people, many of whom volunteered their time. The planning involved the communities at every step, including decisions on where nurseries would be built and the development of land management practices.

The deep understanding of sustainable land management that exists within these communities draws on the wisdom of their ancestors. The Inca and the cultures that came before them were masters of creating high-altitude productive landscapes. They sustainably fed their people with a variety of crops while mitigating erosion, protecting forests, and maintaining soil fertility. The Inca practiced agroforestry and efficiently irrigated and maintained terraced fields. Pollen records indicate that they planted *Alnus acuminata*, for example, a species of alder that can grow at high altitudes and helps stabilize the soil, thereby reducing soil erosion. The Inca also protected woodland resources and managed pasturelands for llama and alpaca herds.

In a recent paper published about the history of environmental stewardship in the Inca Empire, Tino and his coauthors explain that disruption of these practices caused widespread ecological and social decline.[13] The arrival of the Spanish led to the abandonment of the Inca system, resulting in deforestation and the degradation of agricultural land. Acción Andina is reviving the ancient wisdom of sustainable land management and ecosystem restoration techniques, building on cultural traditions that existed across the entire expanse of the former Inca Empire at the height of its power, spanning 2,485 miles (4,000 km) across the continent.

As Florent Kaiser told me recently:

> I see that Indigenous communities, chronically marginalized over centuries, are now turning out to be the essential stewards of the ecosystems we all depend on. High Andean communities produce the water the entire continent will need as glaciers continue to retreat. To me, this reflects an entire sociocultural shift and a fundamental reordering of roles and values. The very communities that were sidelined by dominant systems are now stepping forward as essential guides in facing our shared ecological crisis. Their knowledge, practices, and worldview—long overlooked—are not only relevant but vital to shaping a future where nature and people can thrive together. What's happening in the Andes is not just environmental—it's a cultural reawakening that could reshape how we live, lead, and relate to the Earth.

In the Andes and elsewhere around the world, local communities are restoration pioneers and leaders. But restoring a continent's ecosystems or guaranteeing water security for future generations is a shared responsibility. Governments must step up with public investments and transform policy frameworks that can unlock large-scale nature restoration—and some governments already do. At the same time, the private sector—from hydroelectric energy producers and agricultural industries to banks, insurers, and other entities that benefit from a climate-resilient economy—needs to recognize its dependency on water security and act accordingly. Civil society, too, plays a vital role in expressing the collective demand for a regenerative and equitable future (see Figures 5.2–5.4).

Figure 5.2 Local communities are preparing for a tree-planting festival in the Cusco region of Peru in December 2024.
Credit: Photo by ECOAN

Figure 5.3 Local communities in the Vilcanota region of Peru tend to nurseries of native *Polylepis* trees that will be planted later to restore high-Andean ecosystems.
Credit: Photo by Tuul and Bruno Morandi

Figure 5.4 Participants planting trees in the Aquia region of Peru during a record-breaking event in December 2024. Approximately 150,000 trees were planted in a single day by more than 800 participants.

Credit: Photo by ECOAN

Imagining a Better Tomorrow

Despite early successes, Acción Andina is not a quick fix for the climate or nature. It will probably need to be at least a 100-year endeavor—one that will endure across generations. Continental-scale ecosystem restoration will grow, evolve, and remain alive in the hearts and hands of people for decades to come. No single organization can lead this, but collectively, restoration can be achieved. A century of ecology has begun in the high Andes.

As Florent Kaiser describes:

Acción Andina was never intended as an end in itself but rather a starting and living blueprint—a powerful first step grounded in the region and culture of the Andes. Our aim has always been to work with ancestral principles and community-led wisdom across the world. That's what makes this moment so exciting.

What if Acción Andina is not a unique movement? What if the same reciprocity with each other, our communities, and nature existed in many different parts of the world, waiting to be rediscovered, empowered, and scaled up? If we had more examples like Acción Andina, restoring nature at a planetary scale would no longer be a distant prospect but rather a movement waiting to happen. The good news is that we can find this kind of relationship between local communities and nature all around the world. We only have to look in the right places and with the right mindset.

Living in Harmony with Nature: Satoyama

In 2007, I was recruited by the United Nations Environment Programme to lead the work on forests at the Convention on Biological Diversity. The convention is one of the outcomes of the 1992 Rio Earth Summit and is tasked with no less than the global "conservation of biological diversity, the sustainable use of its components and the fair and equitable sharing of the benefits arising out of the utilization of genetic resources."[14] That is quite a mouthful as a mission. In simpler terms, the convention's objective is to protect the diversity of life on Earth. Over the next five years, I worked on many fascinating political agreements, policies, and standards, including the use of bushmeat in tropical forests and safeguarding Indigenous and local communities from the negative impacts of carbon credit projects, which was a relatively new concept at that time.

One of my most gratifying projects was supporting the establishment of the Satoyama Initiative, an initiative led by the Japanese government. Satoyama landscapes seamlessly integrate villages, farmlands, secondary forests, and grasslands, creating a harmonious blend of human presence and ecological richness. A simple translation of the word *satoyama* as "countryside" fails to capture its profound significance. It is not merely land shaped by human hands but an interwoven tapestry where traditional knowledge and sustainable practices nurture biodiversity and support resilient ecosystems. In other words, it is where humans improve on nature, and nature, in turn, improves human lives. However, Japan's satoyama landscapes are quickly disappearing, as rural populations age and young people move to urban areas. In recent years, there has been a trend in Japan of young people returning to the countryside, although the hard labor and modest pay in agriculture are deterrents to taking up traditional practices. Technology can help bridge that gap and combine rural life with the connectivity and economic opportunities of the digital age.

The notion that nature can enhance our lives is now widely recognized. However, the fact that humans can also improve and help nature thrive is much less well known. Satoyama is a deep dive into our role as ecosystem engineers. One of the landscapes where a stable and healthy relationship between humans and nature is still evident is in Aridagawa, a town in Wakayama Prefecture.[15] This area features terraced rice paddies, a key component of the satoyama landscape. The rice paddies and surrounding secondary woodlands, grasslands, and irrigation systems create a mosaic of ecosystems with a high variety of edge habitats—the areas between two habitats, such as hedgerows and fields. These landscapes are high in biodiversity and are maintained through traditional practices, such as coppicing[16] and grassland management, which have been integral to local agriculture and forestry for centuries. This approach is not unique to Japan. The closer we look around the world, the more it becomes apparent that this harmonious coexistence with nature is more common than we realize, even in the heart of densely populated Europe.

The World's Largest Protected Area Network: Natura 2000

The European Union's nature flagship Natura 2000 is the largest network of protected areas in the world, established in 1992 to safeguard Europe's most valuable and threatened species and habitats.[17] It covers over 18 percent of the EU's land area and about 9 percent of its marine territory, encompassing more than 27,000 sites across 27 EU member states. These sites cover a diverse range of habitats, including forests, grasslands, heathlands, wetlands, and marine ecosystems, and protect over 1,000 priority species, defined as plants or animals in danger of disappearing. The network is not a system of strict nature reserves; instead, most of the areas allow for sustainable human activities while ensuring the conservation of species and habitats through effective management and conservation measures. I started my career as a trainee at the European Commission in Brussels, working for its environment department. My main task at the time was to research and draft a forest management guide for the Natura 2000 network.[18]

Approximately half of the protected areas across the Natura 2000 network are forests and other wooded areas. When researching the origins of these forest landscapes with high biodiversity levels, I was struck by the fact that most of them were not "natural" habitats in the sense of ecosystems without human intervention. Instead, they were mainly forest and woodland areas that humans had managed for centuries or millennia. These ecosystems developed a high and often unique biodiversity, not despite human intervention but because of it.

Examples include coppice forests, where trees are periodically cut back to their rootstocks and allowed to regrow from shoots, creating a dense but relatively low and light habitat crucial for rare bird species like the Eurasian wood grouse, the capercaillie (*Tetrao urogallus*). In other examples, traditional livestock grazing practices combined with large fruit or nut trees have shaped the landscape and biodiversity, requiring continued human management to maintain their ecological value. Examples of these cultural landscapes of high biodiversity include meadows for grazing sheep or cows beneath mature, high-stemmed cherry, pear, apple, or walnut trees. The fruit and nut trees provide additional income and capture more sunlight through photosynthesis, thus increasing overall productivity of the land. More solar energy is converted into growth than could be achieved by grassland alone. The trees also offer shade and shelter to livestock, and their long roots unlock nutrients beyond the reach of grasses and build up carbon deep in the soil. The trees, in combination with the manure from the grazers, provide food and habitat for a wide range of endangered insects and birds, such as the charismatic hoopoe and different types of woodpeckers. Such traditional agroforestry systems reflect historical land-use practices that have shaped the landscape, human societies, and local biodiversity across much of Europe for centuries.

Manage with Nature

When I wrote the first draft of the Natura 2000 management guide for forests, I mainly focused on encouraging human use that had beneficial impacts on forest ecosystems. I recommended certain economic activities in forests as a key component in managing these landscapes sustainably and thought I had done a rather good job. I was therefore taken aback when, during the first stakeholder consultation with the Worldwide Fund for Nature and other groups, there was significant pushback against the active inclusion of private forest owners and forest economic activity in the management of Natura 2000 areas.

The sentiment was that the more we could remove human economic activity from nature, the better. Where necessary, these traditional landscape activities should be replaced by paying some-one to mimic traditional land use, such as grazing with sheep in agroforestry systems. The response from conservation NGOs was more negative than I had anticipated, and it took me some time to understand the underlying mistrust of using economic activity to support nature. It originates from the perception that humans are no longer part of nature, leading to the mistaken conclusion that we must remove the human element from a landscape or marine environment to conserve nature.

However, to the credit of the European Commission and private forest owners across the EU, the management guidelines encouraged the sustainable use of timber and nontimber forest products in most Natura 2000 sites, except in strictly protected areas. When we view ourselves as part of nature and the landscape, we can establish a deeper relationship with nature that extends beyond economic value extraction. Humans can and often do actively benefit biodiversity—after all, we are nature's most powerful ecosystem engineer. We can play that role not only to the detriment but also to the benefit of nature. Let us delve deeper into what a regenerative and rights-based relationship with nature could look like.

The Guardians of Nature

From above, the Amazon rainforest looks like an endless sea of green, a sprawling, dense, and seemingly indomitable force of nature. But the closer you get to the forest on the ground, the more the picture changes. The rainforest is under siege—logging, agriculture, and ille-gal mining carve deep scars into its once-pristine canopy. Yet, in cer-tain regions, the destruction abruptly comes to a halt. Satellite images reveal a stark contrast: On one side, barren land and exposed soil; on the other, a thriving, dense rainforest. The difference? Indigenous stewardship.

I had an opportunity to experience the stark contrast between Indigenous stewardship and deforestation for agricultural expansion during a visit to Ecuador in 2012. At the time, I was working for the UN Programme to Reduce Deforestation and Forest Degradation. We visited a small Indigenous community on their ancestral land in northeastern Ecuador. On their small farm of approximately 40 acres (16 ha), community members were conserving the last remaining forest amid widespread deforestation. Soy and sugarcane fields covered the landscape as far as the eye could see, except for the few thousand rainforest trees around us. Thanks to a little financial support from a government program called Socio Bosque, the Indigenous community could make a modest living from their forest, although in monetary terms, they were earning significantly less than their neighboring farmers from soy, sugarcane, or cattle ranching. But what they lacked in financial income, the Indigenous community made up for in richness of heritage, ideas, and opportunities.

After a lunch of grilled palm weevil larvae, which look and taste like shrimp with a more nutty flavor, we went on a forest walk with our hosts. Our local guides showed us the many medicinal uses of various plants and insects. Trees like *sangre de drago* (dragon's blood) are used for their antibacterial properties, such as healing wounds and treating diarrhea. Other plants are used to treat rheumatoid arthritis, viral infections, and general pain. A plant called *matico* is consumed as tea for respiratory ailments. The roots of the *wasai* plant are used as a diuretic to support kidney health. The list of beneficial plants is vast, and the community's effort to conserve this island of biological and genetic diversity within a landscape of monocultures was far more valuable to society than the small annual subsidy it received from the Socio Bosque program. It felt like we were visiting Noah's Ark, full of nature's ingenuity and traditional ecological knowledge, in an ocean of ignorance and greed.

The Indigenous Effect

Scientists have been measuring the protective effect of Indigenous communities on nature for years, and the results are astonishing. In Brazil, deforestation rates in some Indigenous-managed forests are 22 times lower than in surrounding areas. In Guatemala's Petén region, the rate is 20 times lower. And in some areas of Mexico's Yucatán, deforestation is 350 times lower. It's as if these communities possess an invisible force field, shielding their lands from the destruction encroaching on them.[19]

The effectiveness of Indigenous stewardship is not an anomaly. A study analyzing geographical areas across Canada, Brazil, and Australia revealed a compelling pattern: Indigenous-managed lands are consistently more biodiverse than even government-protected areas.[20] These territories shelter a greater number of species, protect more forests, and store more carbon. The findings challenge long-held assumptions about conservation, suggesting that the most effective environmental protectors may not be government agencies or international NGOs but rather the people who have lived with the land for centuries.

The effectiveness of Indigenous land management is not simply due to ownership. Rather, it is about relationships. Indigenous peoples and local communities manage at least one-quarter of the world's lands, including 28 percent of the forests in Africa, Asia, and Latin America.[21] Unlike conservation models that impose top-down restrictions, Indigenous approaches prioritize coexistence, adaptation, and sustainable use of nature. Forests are not locked away in the name of protection; they are actively lived in, nurtured, and understood. Humans are part of nature, and therefore, nature is more effectively protected.

Lessons from the Land

Indigenous peoples and the land they manage are essential for reaching global forest and nature conservation goals. Across the tropics, except for some areas in Africa, deforestation rates are lower in Indigenous lands than in nonprotected areas and sometimes even lower in not formally protected Indigenous lands than in formally protected areas, underscoring the need to strengthen Indigenous people's rights and recognize their contributions.[22] A study spanning 11 countries from 1990 to 2010 found that expanding Indigenous land rights had a direct impact on forest conservation, reducing deforestation and ensuring healthier ecosystems.[23]

These findings resonate across the globe. In Papua New Guinea, local women act as guardians of the Mangroves, managing and protecting these crucial coastal ecosystems. In Finland, the Skolt Sámi people combine their in-depth knowledge of the land with modern scientific techniques to restore peatlands, yielding both environmental and economic benefits. And in Madagascar's Northern Highlands, the Andrafainkona community has balanced conservation and livelihoods by implementing quotas, issuing logging permits, and supporting community rangers who safeguard their forests.[24]

These stories reveal a crucial truth: Conservation does not require removing people from nature. On the contrary, the most enduring conservation efforts are those led by communities that see the land not as a resource to be extracted but as an extension of their identity. Traditional ecological knowledge, honed over generations, enables these communities to anticipate changes, adapt to challenges, and maintain balance in ways that modern conservation efforts often struggle to achieve.

As the world grapples with the crises of climate change and biodiversity loss, the question is no longer whether Indigenous-led conservation works. The evidence is overwhelming. The question is whether policymakers, conservationists, and global leaders will recognize this reality and support the people who have been safeguarding nature all along. The guardians of nature are already here. All we have to do is open the doors and listen to their stories and solutions. And then act on a new shared understanding.

In pursuing a reconciliation between Western and Indigenous worldviews, it is important not to romanticize Indigenous lifestyles. Many Indigenous people have modern ways of living and communicating; what sets them apart is the deep, reciprocal relationship they have with nature and their community. This deep relationship is not exclusive to Indigenous peoples. We can all reconnect deeply with nature again and build a meaningful relationship with a specific place over time. Some Indigenous peoples and local communities have lived in the same landscape for thousands of years. Nature and people have merged into a new entity known as a landscape.

Respect for Mother Nature

This is a good place to examine Mother Nature's characteristics in more depth. If we form a relationship with her, it is good to know who we are dealing with beyond the facts described earlier in the "Imagining a Better Tomorrow" section. Nature is, of course, not a human being, and our range of human thoughts, emotions, and interpersonal communication is insufficient to describe all of nature. But it is still useful to add a few observations, with a view to managing expectations. Speaking of "Mother Nature" can evoke the image of a benign, always loving and kind, nurturing, and caring entity. I don't want to anthropomorphize nature in this book, so let me add a word of caution about this gentle image. There are aspects of nature we can easily relate to. Often one can find in nature the same broad spectrum of emotions and interactions that make human existence so diverse and often exciting. Nature, just like life itself, has a sense of humor and deals in all aspects of joy, drama, action, sadness, and even irony. Animals, plants, and entire ecosystems adapt, struggle, and thrive in ways that sometimes mirror human experiences, making the natural world a fascinating treasure trove of nonhuman stories waiting to be observed and appreciated. However, nature is not always kind, at least not in the human sense of being kind to an individual.

A significant difference in how humans interpret life and how nature expresses life is that death is an accepted part of the life cycle in nature, while we tend to separate life and death very strictly in our worldview. Humans have a deep-rooted awe and a mystical fear of death, whereas in nature, it is an everyday occurrence that is often necessary to create new life. In other words, death is not as significant in nature as it is for humans. Of course, any living being would rather live than die.

But nature deals out death with seeming indifference to the fate of the individual, and it does so as frequently and freely as it creates new life. Death is a necessary step for life, and the cycle from life to death and back repeats itself endlessly. Our fellow species are used to it and accept it. And often even need it. For instance, some pine trees have seeds sealed in resin, which melts when exposed to high temperatures, allowing the seeds to germinate and sprout. The destructive force of a forest fire ultimately leads to new life and renewal in the forest ecosystem. Perhaps the closer we get to nature again, the more we will lose some of our fear of death. Instead of the end of all things, we will understand it as a necessary step in our evolution.

The Force of Nature

Another common misconception about Mother Nature is that she is weak. In fact, nature could overpower humanity any day. And perhaps someday she will. Nature is, of course, neither male nor female. As I described earlier, I am trying to avoid the pronoun "it," which would signal that nature is a thing or a commodity, which nature is not.[25]

Some believe nature is weak because, mainly since the Industrial Revolution, we have forcefully asserted our dominance over animal and plant populations and often over entire habitats or landscapes. We can easily bulldoze a forest or a wetland to make way for a shopping mall. The forest does not fight back—or, if it does, it may be in the form of protesting youth, other humans winning a court battle on its behalf, or a landslide from deforested mountains. Individual species and ecosystems might be weaker than we are, but nature as a whole is vastly more powerful than all of humanity with all our technology. And whenever we encounter that forceful side of nature, we often fail to recognize it for what it is, because over the past two centuries or more, we have become accustomed to viewing a tamed, domesticated nature that holds no great surprises and can no longer harm us. That is a dangerous misperception.

In reality, nature is so powerful that a tiny virus can bring human civilization to its knees within months. Nature is so powerful that much of a city the size of Los Angeles can burn to ashes before our eyes, without us being able to stop the wall of fire. Nature is so powerful that if sea levels rise dramatically over the next few decades, as some climate tipping points suggest, we will be forced to abandon New York, Mumbai, Bangkok, Shanghai, and many others. No level of human technological effort is strong enough to match the full force of nature. It is a contest we cannot win, not only because nature is stronger but also because nature has almost endless time and patience, while we are quickly running out of time to keep our civilization intact. Our only option is to make peace with nature before it is too late. In fact, we can offer more than a truce. To achieve lasting peace, we can actively restore nature on a planetary scale. Although this sounds daunting, it is no more daunting than other feats of human ingenuity and willpower, such as the moon landing.

Human Rights and Nature's Rights

Once a growing number of us build a new and deeper relationship with nature, a new, mutually beneficial, and regenerative coexistence of humans in harmony with nature can emerge, which we will need to survive and thrive. The alternative is unthinkable: We cannot allow runaway climate change, the sixth mass extinction crisis, and pollution to erode the foundations of life on Earth.

For this new relationship, it is not enough for us simply to invest more financial resources in nature. We can and should also give nature respect and, ultimately, recognize that nature has rights. Let us examine the journey of human rights briefly.

For much of history, forced labor has driven a significant portion of the world's agriculture and economy. Whether it was the serfs of medieval Europe, bound to the land and mainly without rights in a feudal system, or enslaved people working in ancient Rome and Greece or more recently on the plantations and cotton fields of the Americas: It took us too long to overcome the historic wrong of treating a large part of the population as a commodity. Humans are more than things. We have inalienable rights, and we are created as equals. Acknowledging our equality enables us to enter into meaningful relationships with one another, despite all our differences. Universal human rights are, of course, still more an ideal than a reality. Inequality, racism, sexism, and other forms of discrimination are still widespread.

However, human rights are an ideal that enough people believe in and strive for, and therefore they have become part of the moral code of most communities and nations. In ancient Rome, giving an enslaved person better accommodations or food would not have been enough to enable a meaningful relationship. Although those actions are necessary starts, meaningful relationships require respect and rights. It is the same with nature. To build a new and mutually beneficial relationship, we need to respect Mother Nature and her fundamental rights. These rights will differ from human rights, and we need to learn more about what these rights are, how we can identify them, and how we can uphold them. Several jurisdictions worldwide have begun to do so.

The River's Human Face

There is a growing global movement to recognize the rights of nature. New Zealand became the first country in the world to officially recognize a river as a living entity. A 2017 act of Parliament identified the Whanganui River as "an indivisible and living whole, . . . from the mountains to the sea, incorporating all its physical and metaphysical elements." The river has "all the rights, duties, and liabilities of a legal person."[26] To enable the river to assert its rights, two human trustees are appointed in the "Te Pou Tupua": the office of the river's human face. One trustee is appointed by the government, and the other is appointed by Whanganui iwi, the Māori tribes and subtribes associated with the Whanganui River.

In recent years, the rights-of-nature movement has picked up pace, with over 150 laws enacted worldwide as of 2025.[27] Ecuador even enshrined the rights of Pachamama, Mother Earth, in its constitution. This shift to protect ecosystems by granting them legal rights allows them to be represented in court and safeguarded against environmental degradation. For Indigenous communities and environmental activists, the movement goes beyond being a new tool for conserving and restoring essential ecosystems. They view legal rights as a means of respecting nature in a manner that reflects Indigenous worldviews.

Failure to recognize the intrinsic value of nature and related rights carries considerable risk. In his 1944 book *The Great Transformation*, Karl Polanyi outlines some of the reasons why the world economy collapsed in the 1930s.[28] One of the main reasons is that the relatively new global market economy at the time treated all goods and services within the economy as commodities. A commodity is, broadly speaking, a basic raw material or good that is produced to be traded, bought, or sold on the market. A market is a place where buyers and sellers meet and negotiate to establish a price. However, three things that are central to any modern economy are, in fact, not commodities at all: people, money, and nature. Assuming that people (or, rather, their labor) are merely commodities means that our entire human existence and our society are subservient to the economy and the marketplace.

We can all agree that human existence and our civilization encompass realities and experiences beyond being traded, bought, or sold. Humans are not a commodity. Money is not either. If money were a commodity, it could be printed at will until demand is met. And if nature were a commodity, more nature simply could be produced if we run out. None of this is true. And even though, since 1944, we have done a lot to protect people from being traded, bought, or sold as commodities; money from being printed at will; and nature from being parceled out and sold off without a thought to its limited supply or intrinsic value, we are still too often caught in the mistaken assumption that humans, our money, and the nature all around us are mere commodities. For Indigenous peoples, nature is not their external environment, and nature is certainly not a commodity. It is time to move on from that fundamental misunderstanding. New economic frameworks, such as that described in Kate Raworth's *Doughnut Economics*, which make human well-being and a well-functioning society the goal of a modern market economy, are helping to give us a more sophisticated understanding of our role, and nature's role, vis-à-vis the economy.[29]

A New Relationship

So far in this book, we have explored nature as something within us and something that we are part of instead of something external to us. Seeing ourselves as part of nature is a gradual process, and it starts with the first step on the journey: acknowledging that our lives depend on nature. In 2023, Unai Pascual, a researcher from the Centre for Development and Environment at the University of Bern in Switzerland, analyzed more than 50,000 scientific publications, policy documents, and Indigenous and local knowledge sources to gain insights into the role of our values related to nature in policy-making and other decisions.[30] In their work for the Intergovernmental Platform on Biodiversity and Ecosystem Services, Pascual and his coauthors propose a framework of four different stages of our closeness with nature:

1. "Living from" nature emphasizes instrumental values, such as nature's capacity to provide resources for sustaining livelihoods.

2. "Living in" nature focuses on how people recognize nature's importance as settings for their lives, practices, and cultures, particularly in supporting relational values.

3. "Living with" nature centers on nature's life-supporting processes and connections to other-than-human beings, prioritizing intrinsic and relational values.

4. "Living as" nature prioritizes embodying and perceiving nature as a physical, mental, and spiritual aspect of oneself, emphasizing broad values of oneness, kinship, and interdependence.

Different life frames are expressed in varying combinations across time and contexts, but research and policy most frequently align with the concept of "living from nature."

Over time, more of us might come to live "as nature," but for the time being, it would be sufficient if a critical mass of people around the world at least acknowledge that we live "from nature." When I speak about a new relationship with nature, I mean that quite literally, not just as an intellectual exercise. I would like to invite you, and all of us, to build a living, dynamic relationship with nature. And as all good relationships do, it starts with mutual respect. Nature respects humans, at least when we know how to behave in nature. But do we respect nature? Often reciprocity is missing in our relationship with nature. Members of many Native American nations used to carry a small bag of tobacco with them, so they always would be ready to give some back to nature as a token of gratitude and respect. I am not asking you to start carrying a bag of tobacco and drizzle some onto every field, city park, or forest floor. What I do mean is that we can and should all give something back to nature that is valuable to us.

For most of us, usually that is time. Consider giving back an hour a week to nature, by making it a point to go outside, breathe fresh air, enjoy the sky, the wind, the sun, the rain, and whatever variety of plants and animals you can view, touch, or yes, eat: Try exploring the delicious world of herbs and spices that are nowadays considered weeds, such as a dandelion salad or a stinging nettle soup. Take time for a walk in nature, and, if you dare, walk barefoot across the lawn in a park, in the forest, or on a beach. These small steps will initiate a fascinating and ever-deepening relationship with the land, the soil, the elements, plants, animals, and the processes that govern all life. The most effective way to establish a new relationship is to be empathetic and open-minded. If we really want to understand nature, not only from books, the internet, or our favorite AI tool, let's ask nature directly. Let us observe, listen, and ask questions. Be curious. Nature will answer. It may take some time— it's certainly an ongoing process and a learning journey for me.

Perhaps the step that will be most difficult is recognizing that we cannot build a relationship with nature by intellectual exploration alone. We need to build this relationship on an emotional level and even on a spiritual one. In our secular world, where we are used to having all problems solved (or not) by the intellect and by science, asking you to take this step is difficult. But taking this step is worth it, and spending time in nature and with nature will prove it. In the meantime, spending more time in nature is at least certain to be excellent for your health.

In the next two chapters, we examine several projects where the scope and ambition of nature restoration have the potential to reach national or even continental scale.

Chapter 6

A Trillion Trees

We live by the trees. Everything is done by the trees.
—Darius Kamusiime, Farmer, Uganda[1]

Forests have always held a unique place in humanity's relationship with nature. They are with us literally from the cradle to the grave, providing us with shelter, water, warmth, food, medicine, fiber, recreation, and timber. No wonder they are the stuff of fairy tales and myths. One of the world's oldest known epics includes an episode where Gilgamesh, a man portrayed as having superhuman strength searching for immortality, journeys with his friend Enkidu to a vast cedar forest, slays its spirit guardian, Humbaba, and cuts down all sacred cedar trees.[2] After Gilgamesh kills Humbaba and fells the cedars, Enkidu, who was once a creature of the wild, becomes deeply troubled by their actions. He tells Gilgamesh, "My friend, we have made the Forest a wasteland!" (V:303). This comment plunges Gilgamesh into deep despair, and ultimately he is refused his main desire and nature's deepest secret: immortality.

The epic, which emerged in the historical context of the expansion of ancient Mesopotamian civilization into previously unexplored wilderness areas, is one of the earliest cautionary tales against damaging humanity's relationship with forests. It seems to confirm the aphorism "Forests precede civilizations, and deserts follow them."[3] However, other, more positive stories exist of forests and humans entering a symbiotic and regenerative relationship. In this chapter, let's go on a walk into the forest and examine how our understanding and appreciation of these ecosystems are changing and evolving— knowledge we need to save the forests and ourselves.

The Power of Trees

Why are forests so essential for humanity's past, present, and future? In addition to their benefits of stabilizing both the global and the local climate, forests are home to an estimated 300 million Indigenous peoples worldwide, and they harbor most of the world's terrestrial biodiversity. Almost 2 billion people depend directly on forests for their livelihoods. Wood provides excellent material for construction, furniture, packaging, clothing, and most of what a modern bio-economy needs—and, of course, for the irreplaceable coziness and meditative trance that burning logs in a fireplace or a campfire can generate. Forests also provide millions of jobs worldwide, and forest restoration efforts have proven to be a cost-effective way to create new jobs. Every million USD invested in reforestation and sustainable forest management in the United States, for example, can support nearly 40 jobs, including foresters, botanists, technicians, and laborers.[4]

Beyond providing for our livelihoods and daily comforts, forests are vital for the functioning of the entire biosphere. It is often said that forests are the lungs of the world. In fact, much of the oxygen in our atmosphere has accumulated from the photosynthesis of ocean plankton over hundreds of millions of years, bringing the levels of oxygen in our atmosphere to approximately 21 percent and allowing humans and other mammals and animals to breathe easily. Compared to the buildup of oxygen over eons, the forests that exist on Earth today play a relatively small role in maintaining atmospheric oxygen levels. However, they play a crucial role in maintaining the functioning of the global water and carbon cycle, two of the most essential processes for life on Earth. The role of forests in filtering, storing, releasing, and recycling water is unparalleled among other terrestrial ecosystems within the global freshwater and carbon cycle. For instance, forest watersheds deliver clean drinking water for over one-third of all major cities in the world. And forests store about as much carbon as the entire atmosphere. We rely on healthy and intact forests for a stable climate, our water security, and, as we discussed earlier, even the continuity of rainfall over much of our continents. Based on the importance of forests for the water cycle, it is more accurate to call them the world's beating heart rather than its lungs.

Tree Huggers

The important ecosystem services that forests provide are powerful arguments for protecting them. However, there exist even more profound reasons for us to save the world's forests. Just like many of my peers working for nature, I have been focused my career on the conservation and restoration of forests for almost 30 years simply because I am an unapologetic tree hugger: I love forests. I love the smell of forest soil and the shade and coolness of a forest on a hot summer day. I could spend days listening to the sounds of birds, insects, and elusive forest mammals and observing the amazingly complex biodiversity and the miraculous cycle of life that unfolds year-round in forest ecosystems.

Being a staunch supporter of forests and trees does not mean I oppose the economic use of trees. I also love working with wood, and we use it to heat our home. Trees are a wonderful renewable resource: When we manage forests well and for the long term, they can provide us with a perpetual supply of all the forest-sourced goods and services we have come to appreciate. We can protect those forests that need protection, restore those that need restoration, and manage and harvest those forests and trees that can be sustainably harvested. My love of forests and my desire to get to know them better began early in my life, during forest walks with my grandfather, who was a forest engineer for the state government.

During high school, I rallied other students to protect and restore our local forest. Later, during my university studies, I served as the president of the International Forestry Students Association, bringing a youthful perspective into global policy processes and treaties, such as the UN Framework Convention on Climate Change. Forests have been my calling since childhood. I always wanted to work with them and for them. Recently, global movements to protect and restore forests have caught the attention of global companies, philanthropies, and investors. That is very encouraging. And it is about time.

Resetting Our Relationship with Forests

In December 2019, I was on a train journey from Paris to Geneva with my friend Florent Kaiser. We were on our way to a workshop jointly hosted by the World Economic Forum, the United Nations Environment Programme (UNEP), and Salesforce. The meeting's objective was to discuss a bold new plan: the Trillion Trees Initiative, a new, private-sector–driven effort to conserve and restore the world's forests. At the time, I was the head of the Nature for Climate Team at the UNEP's global headquarters in Nairobi, responsible for establishing the UN Decade on Ecosystem Restoration (2021–2030). The proclamation of the UN Decade by the UN General Assembly in 2019 marked a significant milestone in my career, one that I, along with many others, had been preparing for over the preceding year. Before that, I had worked in many countries for almost two decades, trying to save and restore nature and forests. Although I loved my work, it had often felt like an underappreciated endeavor, undertaken by a handful of committed experts and activists—a small world of tree huggers against a vast world of indifference. I was intrigued, therefore, when the World Economic Forum invited me to advise on a new initiative to conserve and restore 1 trillion trees, supported by several multibillion-dollar companies and with the potential to grow into a global movement. It signaled that our small group of scientists and activists, on a mission to save forests, could contribute to building a broad global movement.

The Trillion Trees Initiative

The Trillion Trees Initiative was the bold vision of Marc Benioff—a pioneering tech leader and the CEO of Salesforce. Marc had been inspired to launch the initiative by a 2019 article in the journal *Science*.[5] The findings of that article were subsequently confirmed by a 2023 publication in the journal *Nature*.[6] The milestone *Nature* publication compiled unprecedented amounts of forest data from over 200 leading ecologists and climate scientists and states that forests could draw down 226 gigatons of carbon from the atmosphere, equivalent to over 20 years of total global carbon dioxide (CO_2) emissions, if we optimized the conservation, restoration, and sustainable use of forests worldwide. The *Science* article also highlights opportunities to establish or restore forests covering 350 million square miles (0.9 billion ha), approximately three times the size of India—an enormous opportunity and a daunting task. However, fortunately Marc Benioff is not easily daunted and likes to rise to a challenge. The task of our December 2019 workshop in Geneva was to help prepare the Trillion Trees Initiative for its imminent launch, scheduled for January 2020, at the World Economic Forum in Davos.

The Roots of a Trillion Trees

The current movement to restore forests at global scale goes back to 1977, when the Green Belt Movement began planting trees in East Africa, led by the legendary Wangarĩ Maathai, a human rights activist and environmental campaigner. She defended the Nairobi city forest, Karura, against illegal land grabbing and development, for which she was jailed for six months. But detaining her could not stop the momentum for forests. Her movement grew despite government interference, and the protests against the disappearance of Karura Forest persisted. Eventually, the government relented, released Maathai, and safeguarded the forest. Wangarī Maathai was a pioneer in many ways. She was the first woman in East Africa to earn a doctorate and the first to lead a university department. In 2003, Maathai became the deputy minister of the Environment of Kenya, and she was awarded the Nobel Peace Prize in 2004. And in 2006, UNEP launched the Global Billion Tree Campaign, inspired by Wangarī Maathai's movement. Surpassing all expectations, UNEP reached its 1 billion tree goal just one year later, and then raised the bar to 7 billion trees, which was surpassed in 2009.

Figure 6.1   Plant-for-the-Planet's global campaign with environmental activist and Nobel Peace Prize winner Wangarī Maathai in 2011.
Credit: Photo by Plant-for-the-Planet

Around the same time, a young German activist named Felix Finkbeiner was garnering global media attention with a campaign called Stop Talking, Start Planting! He and his fellow youth activists would search out celebrities and be photographed with a leaf-shaped campaign banner over the celebrities' mouths (see Figure 6.1). His campaign eventually reached the United Nations: At the age of 13, Felix Finkbeiner spoke in front of the UN General Assembly in January 2011, becoming one of the youngest speakers ever to do so (see Figure 6.2). That same year, UNEP's executive director, Achim Steiner, transferred UNEP's Billion Trees Campaign to Felix Finkbeiner's youth movement, Plant-for-the-Planet. Today, Plant-for-the-Planet lists over 300 projects worldwide where forests are being restored. The group also has trained more than 100,000 children and youth from 76 countries as climate justice ambassadors through Plant-for-the-Planet academies. Then, in January 2020, the World Economic Forum together with Marc Benioff and Salesforce took the global

Figure 6.2 Felix Finkbeiner addresses the UN General Assembly in 2011 at age 13.
Credit: Photo by Plant-for-the-Planet

forest movement to a whole new level of ambition by launching the Trillion Trees Initiative as part of the UN Decade on Ecosystem Restoration. The goal was to unite global efforts to grow, restore, and conserve a trillion trees by 2030. The initiative quickly gathered support around the world and since 2020 has gathered ambitious pledges from governments and companies for the conservation and restoration of over 130 billion trees worldwide.

One trillion is an enormous number: 1,000 billion trees would add about 30 percent to all existing forests and trees worldwide. Before the Agricultural Revolution, there were approximately 6 trillion trees on Planet Earth, and we have cut our way down to about 3 trillion.[7] However, it is essential to note that the goal of the Trillion Trees Initiative is not to plant all these trees. Many of the trees that are the focus of the initiative are already standing but are at risk of being cut down and urgently require conservation. The goal of the initiative is to protect, restore, and sustainably manage forests worldwide through new investments, partnerships, and activities that will ultimately affect 1 trillion trees.

Nature's Avengers

What would it take to conserve, restore, and grow 1 trillion trees? Imagine for a moment that you are your favorite superhero with superhuman strength. Imagine you could plant one tree every second, day and night, without stopping. At the pace of 1 tree per second, it would take you 11 days to plant 1 million trees. And it would take 31 years to plant 1 billion trees—a timeline that still fits within a lifetime. However, reaching 1 trillion trees at this pace would take a single person 31,000 years. The Trillion Trees Initiative sets a wildly ambitious goal that no organization can achieve on its own. It was clear from the beginning that the Trillion Trees Initiative would have to work in global partnerships and link up with many other efforts to reach its goal. That is why we met in Geneva to start the initiative as a movement inspired by Marc Benioff's vision. Marc had already convinced other funders, investors, company executives, and the World Economic Forum to come on board with the idea and was now pushing all of us to expedite the rollout toward a launch date. Though we were daunted by the large number we were aiming for, sometimes you have to aim for the stars.

Considering the scale of 1 trillion, it is clear that the initiative cannot be about humans planting and growing all these trees. Given the high costs of tree seedlings, labor, and land, early on we recognized that we would need the most powerful and tireless ally imaginable to grow all these additional trees: Mother Nature. Tree seeds can lie dormant in the soil for centuries, and animals act as natural seed dispersers. Therefore, in most places, there are sufficient tree seeds for forests to regenerate naturally. In many cases, though with notable exceptions we explore later, all we need to do is create natural conditions for restoration, such as by protecting degraded forests, and allow nature to work her magic. Activities pledged by participating governments and companies since 2020 encompass the entire range of conservation, restoration, and sustainable management of forest landscapes. One activity, for example, concentrates on the world's largest protected area corridor in the Democratic Republic of the Congo; other activities support the stewardship and sustainable use of forests by local communities and Indigenous peoples and establish tree nurseries, education, and capacity-building initiatives. Overall, the ambition of the Trillion Trees Initiative is much needed, as forests continue to be under pressure worldwide.

In parallel to planning large-scale forest landscape restoration, stopping deforestation is a top priority for the Trillion Trees Initiative. Even though the pace of tropical deforestation has slowed in the past decade, the world is still losing an estimated 1.5 billion trees across 9.1 million acres (3.7 million ha) of primary tropical forests per year—the equivalent of about 10 football fields of forests being cut down every minute.[8] If deforestation were a country, it would be the third-largest emitter of greenhouse gases, after China and the United States. Conservation of nature alone will not be enough to halt and reverse the loss of biodiversity. We also need to restore part of what we have lost, and that means establishing new forests and regenerating deforested or degraded forest areas. People are the most essential element in all this, and the most critical key to success.

A Rights-Based Approach

Forests rely on the social constructs of human societies for their survival, including economic drivers, ownership structures, governance, and other pressures and triggers of deforestation or forest restoration. The high reliance on wood fuel for cooking across Africa, for example, is a major driver of forest loss and forest degradation. This problem could be addressed by switching to clean cookstoves where possible; where charcoal is still the fuel of choice or necessity, establishing woodlots and more trees on farms could provide a sustainable supply of wood fuel. One could say that the challenge of ecosystem restoration is primarily about understanding and balancing the needs, aspirations, and motivations of people and communities across a landscape. That is why a rights-based approach that focuses on the needs of local communities and entire societies is essential for successful forest conservation and restoration.

The Trillion Trees Initiative aligned itself from the outset with the UN Decade on Ecosystem Restoration 2021–2030 with its overall objective to "prevent, halt and reverse the degradation of ecosystems worldwide."[9] The UN employs a rights-based approach to its projects and programs, grounded in the Universal Declaration of Human Rights and other documents, such as the UN Declaration on the Rights of Indigenous Peoples. Principles such as "Free, Prior, and Informed Consent" of Indigenous and local communities regarding any project that impacts their rights have been enshrined in international law, as seen in the UN Framework Convention on Climate Change. For ecosystem restoration, the UN Decade outlines 10 principles for good practices that all projects associated with the Decade should follow, including projects under the Trillion Trees Initiative.[10]

In summary, the 10 principles emphasize aligning efforts with global policy goals to reverse ecosystem degradation, improving overall ecosystem health and functionality, and protecting native biodiversity. They aim to improve overall human well-being through restoration, inclusive engagement of diverse stakeholders, and integration of various knowledge systems. They advocate for setting measurable goals, prioritizing native species, and considering the broader landscape context. Finally, the principles emphasize aligning restoration efforts with existing national and local policies to ensure long-term effectiveness and replicability.

Together, these principles guide restoration projects across different scales, fostering a holistic approach to ecosystem recovery. The Trillion Trees Initiative has subscribed to those 10 principles and expects all its partners to uphold them.[11] Implementation and monitoring are the responsibility of individual member companies, and the World Economic Forum aggregates the company reports on an annual basis. Although few projects accomplish all these things flawlessly, most are well designed and well implemented, and they continue to improve as they mature, as tools and technologies evolve, and as capacity increases. Most companies rely on highly experienced nongovernmental organizations for the implementation of their pledges, such as American Forests, Arbor Day Foundation, Conservation International, Plant-for-the-Planet, The Nature Conservancy, and WeForest, to name just a few.

Perfect as the Enemy of Good

Striving for perfection is necessary when it comes to ecosystem restoration, but we must also be pragmatic and recognize that we need both speed and scale in the present crisis, which means learning by doing. Each local project context is different and might require tailor-made solutions. For example, the Karura Forest in Nairobi was once unsafe to walk in due to rampant crime, was littered with plastic and other garbage, and native trees were being cut down for firewood. Fencing in the forest and collecting an entrance fee from visitors initially upset some local community members, but it was necessary to protect and restore the forest. The proceeds have been used to make the forest safer and cleaner and to remove invasive eucalyptus trees. Although the fencing and fees have curtailed access for some, they are essential for making the forest the safe and the favorite weekend and leisure destination for many Nairobi citizens that it is today. Karura Forest also now is a regular spot for school outings, where classes enjoy guided walks or bike rides in the restored forest with beautiful native trees.

Some early restoration activities of Karura Forest, such as its fencing, seem at odds with the principles of prioritizing local community rights. But in the end, all of Nairobi is now enjoying a healthier, more prosperous, and better-managed protected forest in the heart of the city rather than an open-access area that was described previously as a den of thieves and an open garbage dump and was slowly being denuded of its trees. Some access rights had to be limited in the short term, but conserving and growing tens of thousands of trees within the city boundary benefits everyone today. The forest helps secure Nairobi's drinking water, purifies the air—vital in a city with serious air quality challenges—and can bring Nairobi's 4 million citizens closer to nature.

Global initiatives like the Trillion Trees can learn from the hundreds of thousands of local initiatives, such as the restoration of Karura Forest, that are based on decades of experience in conserving and restoring forests. Fortunately, finding and amplifying this knowledge is becoming easier than ever, thanks to artificial intelligence that can track, translate, and transmit the learnings from success stories and failures to an increasingly interested and growing community of forest investors and project developers. Over time, one goal of the World Economic Forum and the Trillion Trees initiative is to use the power of AI to enable joint learning and knowledge sharing from the Trillion Trees Initiative experience across all its participating organizations and other major restoration platforms, including the UN Decade on Ecosystem Restoration, the project platform Restor, and the Global Mangrove Breakthrough.

Trailblazer

Two years after the launch of the Trillion Trees Initiative, I was approached by Salesforce asking me to lead its forest and nature activities and investments, including its contributions toward the Trillion Trees Initiative. I thought long and hard about the offer, not only because I loved my job at the UNEP, which included running a fantastic global team and a program for the conservation and protection of forests under the Paris Climate Agreement. I was also generally skeptical of the private sector. There was too much greenwashing and too little understanding of the need for systemic change—at least, that was my initial bias. However, after meeting the team at Salesforce, exploring the company's track record, and reading Marc Benioff's book *Trailblazer: The Power of Business as the Greatest Platform for Change*,[12] I was deeply impressed with their sincerity and the company's potential to trigger change across the world's entire private sector. I joined Salesforce in May 2022 as Vice President for Climate Action, working with a small team on nature and water impacts and investments, such as the Trillion Trees Initiative, and with our impressive Philanthropy Team. Since it was founded in 1999, Salesforce has donated almost 1 billion USD for community efforts and global change, with a focus on education and sustainability. The company is headquartered in San Francisco, but I preferred to stay in Europe, a convenient time zone for a globally operating company. I was also asked to coordinate the activities of our sustainability team in the EU, Australia, Brazil, France, India, and Japan.

In *Trailblazer*, Marc describes the powerful ways in which large companies can trigger and lead change for good. And while I believe that the street remains the largest platform for change, business is undoubtedly the fastest and most innovative. With the power of a multinational, multibillion-dollar company comes enormous responsibility. It is commendable that many companies have joined the Trillion Trees Initiative. Their reasons for joining include investing in forests as a pathway to achieving net-zero carbon emissions targets and ensuring the health and resilience of their supply chain, such as coffee, cocoa, other agricultural products, or pharmaceuticals.

Since its inception, the Trillion Trees Initiative has inspired almost 100 companies to collectively pledge the conservation and restoration of almost 10 billion trees. In addition, the governments of the United States, Canada, China, the Democratic Republic of the Congo, and India have pledged to conserve or restore a total of 122 billion trees. Altogether, the Trillion Trees Initiative now aims to conserve or restore approximately 132 billion trees, and further pledges are still being received.

Making pledges is easy, though. The challenging part is implementing these conservation and restoration activities in a manner that benefits people, the climate, and biodiversity. At the beginning of the Trillion Trees Initiative some mistakes were made, such as using the number of trees as the primary metric for success, leading to projects that aimed to plant as many trees as quickly as possible. Some projects that were not successful led to accusations of greenwashing and some pushback from NGOs and the media. We all learned a few lessons the hard way, and, for most companies, the learning journey is steep. However, there is no shortcut to any place worth going to. Saving and restoring the world's forests is undoubtedly worth achieving, despite difficulties and headwinds, so we keep going and keep learning.

Resilient Nature, Resilient Business

The future supply of nature's goods and services is something we tend to take for granted, and we often underestimate the risk of biodiversity loss to our own well-being. Approximately half of all modern pharmaceuticals, for example, originate from natural products, many of which are derived from the biodiversity found in tropical and temperate forests. Ensuring that this vast reservoir of biodiversity, which might hold remedies for cancer or other currently incurable diseases, remains intact is an investment in our collective health and well-being and in the medical supply chains of the future. Forest ecosystems also contribute to our health in other ways, of course. Salesforce was motivated to commit to funding 100 million trees by the health and resilience benefits of trees for the local communities where our more than 70,000 staff live and work. Trees can help mitigate heat waves, floods, and droughts and significantly add to the value of urban settings and rural landscapes.

Another reason for Salesforce's interest in the Trillion Trees Initiative is that, as a global technology company, it relies on data centers for its computing power and reliable data storage. Many data centers use water to cool their vast arrays of servers and processing units. Forests can help make watersheds more resilient and more able to provide sufficient clean water for households, for critical industries like agriculture, and for data centers. Other companies in the Trillion Trees Initiative, such as banks, pension funds, and insurance companies, view forests and nature as an emerging asset class and seek to learn how to invest wisely in forests, meaning with long-term benefits for the climate, nature, and people, in addition to solid financial returns. Saving and restoring the world's forest ecosystems is an idea whose time has come, and we need all key financial and political decision makers as well as all forest stakeholders on board.

The Trillion Trees Initiative was set up to do just that: Leading by example, it aims to inspire thousands of large and medium-size companies to invest in forests and governments to set the enabling policies. Salesforce is particularly well placed to bring other companies along, based on its unique leadership role within the private sector: Among its customers are not only 60,000 NGOs but also tens of thousands of large companies, including most of the top 500 corporations worldwide as measured by revenue. These companies place great trust in Salesforce, which manages and protects a significant portion of their data. Therefore, the sustainability teams from many of the world's largest companies often compare notes on investments and priority activities related to nature, water, and climate action with our team. For Salesforce, it is a position of considerable opportunity and responsibility. Investing in forests for climate, biodiversity, and livelihood benefits is a clear recommendation we give to other companies, and many have followed our call to action. As I explain next, investing well in forests is challenging, but a clear commitment and willingness to learn are good starting points.

A Landscape Approach

One of the activities my team and I are currently responsible for at Salesforce is helping to fund the conservation and restoration of 100 million trees by 2030. As of 2025, we have supported the conservation and restoration of nearly 60 million trees in over 30 projects across 15 countries, with many partner organizations. In 2024, we commissioned an independent evaluation of our 29 largest projects.[13] The team from an independent consultancy utilized high-resolution satellite images, field visits, and expert analysis, supported by AI, to identify areas of progress and potential problem areas. Progress in most of our forest projects was good, though the tree size in some projects was not yet large enough to measure progress conclusively using satellite imagery alone, even though we utilized the highest non-military-grade satellite imagery (at 30-cm resolution). However, we also encountered a big surprise. We were unable to identify a sufficient number of planted trees in one of our most significant projects in Madagascar. Only about one in four of the trees we expected to find were actually there.

We quickly recruited a globally renowned expert based in Madagascar and asked her to conduct an investigation. She spent several days in the field, including a full day of travel to and from the remote project area on an island in a river delta. She found that our local tree-planting partner had not considered the fact that the project area was experiencing high deforestation rates due to the poverty of the local community. Citizens from a nearby town were felling mangrove trees to burn and sell charcoal, and the tree-planting project could not counterbalance the rapid disappearance of trees. In addition, drought and a flash flood had killed many of the young trees. And finally, a less-than-ideal species of mangrove trees had been planted. Although these trees were quick and easy to propagate and plant, they were less likely to survive in the specific local hydrology and climate conditions.

In hindsight, a more effective approach would have been to create alternative income opportunities for the charcoal traders, thereby stopping deforestation and relying on natural regrowth instead of planting new trees. The choice of easy-to-plant mangrove seedlings was likely driven by our initial requirement to plant a large number of trees quickly within a limited budget. Ultimately, these became our most expensive trees because the project failed.

Learning from the Forest

Everyone involved in this project learned the hard way that ecosystem restoration needs a holistic approach at the landscape level. A landscape approach is commonly understood as a planning and implementation approach that considers a large enough area—for example, an entire watershed area or an entire river basin—to understand holistically what drives change across that landscape, centered around the people who live in it. This landscape approach also evaluates the drivers of forest degradation and loss beyond the project area and designs solutions based on the identified socio-economic and ecological context. Rather than focusing on a single sector, such as agriculture or tourism, a landscape approach brings together diverse stakeholders, including farmers, local communities, businesses, and government agencies, to achieve and balance multiple objectives across an entire landscape. This approach had not been used in preparation for the project in Madagascar, and we and our local project partner learned an important lesson. The partner has since revised its entire project development strategy for all its projects worldwide. Project development now starts with examining the landscape holistically in terms of social and ecological context. Based on the factors driving forest degradation, solutions are designed to halt and reverse forest loss while generating benefits for local communities.

Our evaluation also identified highly successful projects that exceeded our initial expectations. One Uganda project, run by the Jane Goodall Institute, employs local farmers to plant and care for trees. When the evaluation team interviewed a farming couple in the project area, they shared that they had started to benefit from the trees by selling branches for firewood, thereby diversifying their farm income. They were now planning to invest their own money in planting more fruit trees as well as more trees for timber and firewood. The shade and organic material from additional trees improved their agricultural yield. Such virtuous cycles—more trees in the landscape lead to more income for farmers and, in turn, to more investments into nature—are becoming more common, as our understanding of the importance of forests and trees across the landscape grows. One result of our evaluation was that Salesforce will make all projects public on the global, AI-driven online platform Restor (www.restor.eco), which already brings together more than 250,000 restoration projects. The platform enables each project to monitor its impacts on soil, forest carbon, biodiversity, water, and other benefits for local communities. It also has an almost real-time fire alert system for forests and other ecosystems worldwide. With this and other technological advances in project planning and monitoring, investments in forest ecosystem restoration can continue to become more effective.

The growing enthusiasm for forests and trees is a good thing. Ecosystem restoration will be critical in turning the tide against climate change and achieving humanity's overall agreed blueprint for a livable future, the United Nations Sustainable Development Goals. But we need to be mindful of some pitfalls that may arise along the way. Over the decades, our global community has learned valuable lessons in afforestation and other restoration projects across dozens of countries and most ecosystems. Best practices that have emerged from these lessons include using a landscape approach rather than focusing on small project areas at a time. Such best practices can help reduce costs and minimize future risk, as the world embraces the need to conserve or plant more trees.[14] Alongside smart and effective project implementation, we also need policy change, such as shifting public incentives toward restoration and conservation and creating predictable income sources from forest products and forest carbon. A valid first rule of restoration is that the conservation of remaining natural areas is the top priority. Removing the drivers of degradation can be the best and cheapest tool to let nature recover. And with the growing enthusiasm of companies and individuals to invest in forests, we often encounter a fundamental misunderstanding: that all new trees must be planted.

To Plant or Not to Plant?

Today we know that most new trees do not need to be planted. Most previously forested ecosystems in the world have remnant seeds in the soil, or birds and other animals acting as seed dispersers that bring new tree seeds to degraded lands at no cost. Natural regrowth often is a more cost-effective and successful alternative to tree planting. The most cost-effective type of restoration is working with the forces of nature while gently guiding and accelerating them. This process is known as assisted natural regeneration.

For instance, farmer-managed natural regeneration is a successful and fast landscape restoration technique across the Sahel. The process begins by identifying remnant rootstocks below the surface, where the trees aboveground were cut years ago but their rootstocks remain alive. Just as in coppice forests discussed in the section on Natura 2000 in Chapter 5, many tree species whose trunks have been cut will remain alive in their roots for decades, and send up new shoots. Farmers can nurture those "underground trees" back to life, by carefully pruning the regrowth to a single stem and protecting it from browsing by goats, sheep, or cattle. By concentrating all the water and energy of its sizable underground root system into one new stem, trees recover to full growth very quickly. The results are stunning—within a few years, large trees cover the once-barren and dry savannah, bringing back water, shade, productivity, and life. In this way, millions of hectares across the Sahel have been restored using not much more than a pruning knife and a new grazing regime for livestock, keeping them away from the trees long enough for them to grow out of reach of animals. In semiarid regions like Niger, trees supported by farmer-managed natural regeneration can reach up to 6.5 feet (2 m) in height in the first year and 10 feet (3 m) or more by the second year. The trees grow so fast, even in this desert-like climate, because they can draw on stored energy reserves in the existing root systems. Pioneers like Australian agronomist Tony Rinaudo, who is widely known in the region as the forest maker, and his organization World Vision have helped to spread this successful technique from farmer to farmer across the Sahel.[15] By now it has reached critical momentum, spreading by word of mouth without much external support.

However, in some areas, forests will not regenerate fast or well enough on their own, and planting tree seeds or seedlings is the only option. The reasons can be, for example, because degradation is so advanced that seedlings cannot take root without irrigation or other assistance, or because any forest remnants are too distant to allow for natural seed dispersal. Often tree planting must be done manually, not by drones or other technologies. In some remote areas, drones are used to fire seed pods into the ground with high efficiency and at low cost, although often at the expense of tree survival rates. The impact of the seed pods further compacts the soil, making it difficult for roots to take hold. The high Andes, for example, lacks remaining mature *Polylepis* trees that can produce seeds for natural restoration, and the surrounding landscape is often so degraded and the soil so compacted that seeds would not germinate without human help. Similarly, in areas devastated by severe forest fires, there may be a scarcity of seeds from well-adapted tree species, making human intervention necessary.

Food for Goats

Before new trees are established, it is essential to address the drivers of degradation, such as overgrazing by cows or goats or deforestation for charcoal or firewood. I have seen poorly planned tree-planting projects that did nothing more than plant expensive goat food: All the planted trees were gone again in under one growing season. Knowing and addressing the local pressures on a landscape is essential. For example, new and more cost-effective tree-planting efforts are being tested in the Brazilian Amazon and in Laos. In both regions, instead of planting tree saplings, tree seeds are sown directly into the soil using mechanized seed dispersal methods usually used for planting grains. This method has been effective at restoring 90 tree species from seeds in sites up to 123 acres (50 ha) in tropical wet forests in the Brazilian Amazon, and mechanized direct seeding costs are less than half that of planting seedlings.[16]

There exists an impressive body of knowledge about which tree species to plant, and when and where to plant them, and of other techniques, such as assisted natural regeneration that rely on natural seed dispersal.[17] Dozens of countries already have detailed maps of where the best restoration opportunities can be found and how to restore forests and landscapes. Typically, indigenous tree species are preferred, but in a rapidly changing climate, it is essential to recognize that the natural ranges of trees are shifting. The Trillion Trees Initiative is building on the lessons learned from all these projects and programs. It has established a world-class scientific and social advisory council to help guide the initiative's efforts. However, even with the best scientific advice, investing in the most effective ways to conserve and restore forest ecosystems remains a learning journey, because no ecosystem is exactly like another, and no local community's or country's needs regarding ecosystem goods and services are exactly alike. Each project is unique and requires careful study, design, implementation, and monitoring. Resetting our relationship with forests takes time, and we need to build trust in our relationship with all the people who rely on forests. Projects that violate that trust undermine the movement as a whole.

Forests Under Scrutiny

The global tree-planting movement has received significant media attention recently, with initial coverage largely positive and enthusiastic. At the beginning of the 2020s, tree planting was at times portrayed as a simple, tangible solution to combat climate change, with ambitious campaigns and initiatives garnering widespread support. For example, the July 12, 2019, cover of *The Guardian Weekly* stated "This Machine Kills CO_2," over a picture of a tree.[18] However, as the movement has grown, so has scrutiny and criticism, particularly regarding challenges related to the often-nuanced and complex implementation of projects on the ground, which always involve numerous stakeholders, typically thousands of people or even entire towns and cities. There has also been some backlash against portraying tree planting as the silver bullet solution against climate change.

Let us be clear: Saving and restoring forests will not, by itself, be enough to save the climate or nature. To limit global warming, we also need to build a circular global economy powered by renewable, clean energy. And to save biodiversity, we need, first and foremost, to change the way we produce food and manage our land and seas. But forests and other ecosystems definitely must be part of the solution for the nature and climate crisis. Turning the tide on forest loss and degradation is complex, and generalizing rules for forest projects to the global level is almost impossible because the conditions of each forest and each project are so different.

Some media coverage has highlighted the complexities and potential pitfalls of forest restoration. Reports have emerged of failed projects, such as in Turkey, where up to 90 percent of saplings died shortly after a mass tree-planting campaign, and in Australia, where extensive land clearing negated the benefits of new plantings and deforestation outpaced restoration efforts. These stories have raised questions about the effectiveness and long-term viability of many tree-planting campaigns.

Critics argue that many companies and organizations use tree planting as a superficial way to appear environmentally responsible while avoiding more substantial changes to reduce their carbon footprint. This suspected greenwashing allows businesses to maintain a facade of environmental action without addressing the root causes of their emissions. And while it is of course true that companies need to do more than plant trees to claim net zero or nature-positive operations, recent studies have found that companies that buy carbon credits, including from nature-based projects, are on average also decarbonizing their operations and value chains twice as fast as companies that do not invest in carbon credits.[19] In other words, it is both essential and possible for companies to invest in nature and climate action through carbon or biodiversity credits while reducing their emissions across their operations simultaneously.

Quality Over Quantity

Poorly planned tree-planting initiatives can have negative consequences, such as the introduction of nonnative species, the creation of flammable monocultures, and the disregard for the rights of local and indigenous communities. Such failed projects have contributed to skepticism about the actual impact of tree-planting campaigns and their potential to address climate change effectively. Problematic projects continue to exist. The majority of the projects we reviewed in our Salesforce portfolio and of the quarter million projects on the online platform Restor.eco, however, are well informed, well implemented, and well monitored. There has been an extremely fast and broad learning and maturation of the global restoration movement in the past few years. The impact of global tree planting and the broader forest conservation and restoration movement is overwhelmingly positive, and it far outweighs any adverse consequences from individual projects that may have made mistakes. Unfortunately, projects with low-quality implementation tend to receive more press coverage, although their numbers are significantly smaller than the majority of well-designed, innovative, inclusive, and socially and environmentally beneficial projects. We dive deeper into some of the world's most successful projects in Chapter 7.

Over the past five years, since the Trillion Trees Initiative was established, the media echo has been swinging between extremes. At times, in particular in the beginning of the initiative, tree planting projects were touted as the silver bullet to solve climate change. At other times, they were supposedly all ineffective greenwashing. Neither is true, of course. The truth is, as so often, more nuanced and much closer to common sense. Trees and forests are not the silver bullet for climate action, but we will reach global climate, biodiversity, and socioeconomic development targets only if forests play a major role. And while tree planting is still a valuable tool in a broader toolbox of forest investments and management activities, there is an increasing emphasis on the need for more holistic approaches, such as forest landscape restoration, that consider overall community well-being and ecosystem health. This evolving narrative underscores the complexity of environmental solutions and the importance of critically evaluating climate mitigation and adaptation strategies. Having worked on this topic for over 25 years, I am impressed with the fast learning pace and the increasing maturity and sophistication of nature-based solution investors and implementers in recent years. Investors in nature and climate action are starting to embrace the complexity of this space, and the 11-fold increase in private-sector investments into forests and other ecosystems between 2020 and 2024 shows that we are now on an exponential solutions curve.

Is It Too Late?

Resetting our relationship with nature and stabilizing the climate is a race against time, and some say we are already doomed. Interestingly, widespread climate change denial seems to have given way to climate "doomism" and apathy about the state of our natural world without consideration of serious, ambitious climate and nature action. In his book *The New Climate War*, Michael E. Mann states that doomism has replaced climate change denial as the most widespread obstacle to building political will.[20] Such doomism is as dangerous for our future as climate change denial, because it robs us of the powerful agency for change that we inherently possess. Reclaiming our agency means taking responsibility. But is it too late?

From a survey of 400 climate scientists in early 2024, *The Guardian* distilled a headline that our climate response is "hopeless and broken."[21] Seventy-seven percent of respondents expected global temperatures to rise by at least 4.5°F (2.5°C) above pre-industrial levels by 2100, which would cause widespread devastation across ecosystems and human societies. The main culprit identified by most scientists is a lack of political will. It sounds like we need a miracle to help our civilization survive. However, in reality, the miracle is already here, in the form of the renewable energy transition, which is gathering speed, mostly due to the fast drop in prices for wind and solar energy and better storage technologies and to our knowledge that reversing nature loss can contribute one-third of the solution to the climate crisis. All we need, in fact, is political will. And "political will is itself a renewable resource," as Al Gore puts it.[22] It can be generated by minting concerns, fears, hopes, and aspirations into concrete policy proposals and investments. It is not too late. However, harnessing nature's healing power to solve the climate crisis is definitely a race against time, because climate change itself is undermining nature.

A Race Against Time

Forests have steadily absorbed a significant portion of the excess carbon dioxide released into the atmosphere over the past few centuries and decades. That they will continue to do so is no longer a given, because carbon storage in forests is not a one-way street. When forests die, decompose, or burn, they emit carbon into the atmosphere. Until now, forests sequester about twice as much CO_2 as they emit. A study in the journal *Nature* found that between 2001 and 2019, forests absorbed a net 7.6 billion metric tons of CO_2 annually.[23] That is more than one and a half times all of the US emissions in 2023, or nearly all the emissions from the global transportation sector combined: all cars, planes, and ships.

However, the speed of carbon absorption by forests around the world shows signs of weakening or even reversing. There is growing evidence that the carbon absorption capacity of forests is declining; and in some cases, forests are transitioning from carbon sinks to carbon sources. Studies show that since 2021, Finnish forests, for example, have transitioned from carbon sinks to carbon sources, emitting 1.12 million tons of CO_2 in 2023.[24] This is happening because a warming climate makes forests more susceptible to wildfires, droughts, insect outbreaks, and disease. If more trees are being cut, or are dying or burning in forest fires, than new trees are growing, a vital climate change buffer will be lost.

Spreading Like Wildfire

The increase in wildfires globally is of particular concern for the future of forests. The risk of new megafires, which can envelop entire cities, as happened in Los Angeles in the state of California in January 2025, is yet another alarm call from nature to ensure our economies and all governments, companies, and citizens to shift to cleaner, renewable energy and reduce emissions as quickly and widely as possible. The increasing risk of forest fires also highlights the urgent need for intensified forest protection, management, and restoration efforts to sustain their vital role as carbon sinks. Although dense forest undergrowth, of course, does not in itself cause forest fires, the risk of fires can be lowered significantly through effective management approaches to mitigate forest diseases and reduce the availability of fuel in the undergrowth.

In 2022, my team at UNEP commissioned a report on the global threat of wildfires. *Spreading like Wildfire—The Rising Threat of Extraordinary Landscape Fires* outlines how the global wildfire crisis is intensifying.[25] The study found that large, hard-to-control fires, known as megafires, are on the rise and could increase up to 50 percent by 2050. Forest fires have now surpassed land clearing for agriculture as the main driver of forest loss. Climate change and land-use changes are primary drivers of this trend, leading to an anticipated increase in extreme fires even in previously unaffected areas. Recent record-breaking wildfire seasons across various regions, including Australia, the Arctic, and the Americas, highlight the urgent need to address this growing threat.

Our team at UNEP and the global Think Tank in Norway, GRID-Arendal, proposed a new "Fire Ready Formula" that emphasizes prevention, preparedness, and recovery.[26] This approach calls for a shift in government strategies, focusing on ecosystem restoration as a key component in minimizing wildfire risks. By implementing better preparation measures and adopting sustainable rebuilding practices in the aftermath of fires, such as fire breaks, early-warning systems, and tree nurseries for sufficient reforestation supplies, communities can enhance their resilience to future wildfire events. This holistic approach aims to reduce the occurrence and impact of extreme wildfires while fostering more fire-resistant landscapes and communities in the face of ongoing climate change. It is encouraging to see that the World Economic Forum, through the Trillion Trees Initiative and other efforts, is giving more attention to this topic. The Trillion Trees Initiative recently launched a global wildfire initiative, joining forces with the UN, governments, and private companies to ensure we can all be more fire ready.

The Trillion Trees Initiative is not the only global effort aimed at boosting the protection and restoration of forests; it is not even the biggest one. Multilateral efforts under the Paris Climate Agreement, for example, aim to increase financial investments into forest protection, and other international funds for forests have been established, such as the new Global Biodiversity Fund under the UN Convention on Biological Diversity. However, Trillion Trees is the largest forest initiative driven by the private sector, with unique advantages and knowledge to bring to the table. We all need to work together—governments, civil society, and the private sector—to fix our relationship crisis with Mother Nature and turn the tide on forest loss. When we work together, extraordinary success stories are possible: planetary-scale ecosystem restoration projects that change the face of landscapes and societies. In Chapter 7, we visit a few projects that aim to heal nature—and ourselves—at the scale of an entire country or even a whole continent.

Chapter 7

World Restoration Flagships

But where there is danger, the saving power also grows.
—Friedrich Hölderlin

In previous chapters, we established that conservation, restoration, and sustainable use of nature are more than just nice to have. Nature is a critical component of the infrastructure that supports human civilization. More importantly, nature is the basis for our mental, spiritual, and social well-being. We have already lost more of nature than we can afford to lose.

How do we begin to plan and implement the restoration of nature on a global scale? At the unprecedented scale of entire cities, large landscapes, countries, and continents? Fortunately, some courageous projects and programs in the world have started this monumental task. A few have already achieved a moonshot ambition level for nature or will soon reach it if we all help. In 2022, the United Nations established a new award for some of the most promising and inspiring large-scale restoration projects: the UN World Restoration Flagships. In this chapter, we visit a few of these lighthouse initiatives around the world. The next one may soon come to a landscape near you.

Brazil's Atlantic Forest

When the first European settlers arrived in 1500 on the northeastern coast of Brazil, they would have found a seemingly impenetrable wall of dense forests stretching for hundreds of miles inland. It was a magnificent forested world, extending from what is today the state of Bahia in northeastern Brazil, southward along the Atlantic coastline, and inland into northeastern Argentina and eastern Paraguay. The Atlantic Forest (Mata Atlântica), a complex forest biome once approximately twice the size of the US state of Texas or nearly half the size of India, comprised many different ecoregions, including tropical and subtropical rainforests, coastal forests, mangroves, and Araucaria moist forests, a rainforest reaching up into mountainous areas with frequent frosts and snow during the winter months, dominated by the charismatic Araucaria tree, also known as monkey puzzle or candelabra tree. This "other rainforest" of South America was indeed a wonder of nature. At the time, it was the third-largest rainforest on the planet, after the Amazon and the Congo Basin. Only about 20 percent of the original Atlantic Forest is left today, and less than 2 percent of the remaining forests are legally protected.

However, contrary to popular belief, in pre-Columbian times this natural wonder was not an unmanaged forest. The Atlantic Forest and similar tropical forest biomes worldwide have been inhabited by Indigenous peoples and cultures who have managed them for millennia using sophisticated ecological knowledge. A recent study from the Amazon Basin, for example, showed that 85 woody species domesticated by pre-Columbian peoples were five times more likely to be abundant near archaeological sites in Amazonia than in locations farther away from historic settlements.[1] In other words, Indigenous peoples selected, cultivated, and nurtured trees with edible fruits, seeds, and plants for medicinal or cultural use on a large scale.

Over thousands of years, these activities would have changed the composition of the forest. This effect is less well studied in the Atlantic Forest because so much of it disappeared before anyone questioned its original relationship to human civilization. However, it is safe to assume that similar forest management patterns and species selection to support human needs occurred across the Atlantic Forest.

A Forest Garden

In the remaining pockets of native Atlantic rainforest, there exists an extraordinary number of tree species with edible fruits, nuts, or leaves and a vast diversity of plants, insects, and fungi with medicinal or cultural uses. It can be said that, in some ways, the Atlantic Forest, like much of the Amazon, was a vast forest garden managed by and for the people. Recent archaeological evidence also suggests that the Amazon basin was densely populated in the pre-Columbian past. Conservative estimates put the population count between 1 and 2 million; according to the highest estimates, 8 and 10 million people lived in the Amazon Basin before European colonization.[2] Recent discoveries of extensive earthworks, settlements, and agricultural areas across this area support the idea of dense pre-Columbian populations. For example, research in Brazil's Upper Tapajós Basin revealed 81 settlements, indicating a higher population density than previously thought.[3]

The recent change in knowledge about Indigenous peoples also extends to other areas of the Americas. The English settlers who arrived in Virginia in North America in 1607 assumed that large parts of the future United States were untouched, pristine wilderness. In reality, much of the entire continent, including forests, grasslands, prairies, and mountainsides, had been expertly managed and shaped for millennia by Indigenous land stewards who used fire as the primary management tool to optimize land for hunting. They periodically burned large, open areas, grasslands, and forest understories, thereby easing travel, driving wildlife into specific areas for hunting, and promoting the fresh growth of food plants, such as berries and nuts.[4] Our emerging understanding of the effectiveness and extent of traditional ecological knowledge and practices—and of the beneficial role of humans in forest landscapes and other ecosystems—challenges the conventional notion that ecological restoration primarily means returning to a state of nature before human intervention. What is actually needed is a shift in our human understanding of our role within nature and our management of ecosystems. From traditional ecological knowledge, we can learn how to return to the abundance and high productivity of diverse forest landscapes. Our new relationship with and understanding of nature, not removing humans from nature, is what will lead to successful conservation and restoration at global scale. Recent guidance for effective ecosystem restoration reflects the essential role of humans as stewards of nature. The principles and standards of the Society for Ecological Restoration, for example, describe restoration as a process that engages stakeholders at every step.[5] And that is essential, because degraded land is usually populated by humans.

A Modern Rainforest Stages a Comeback

Today, the Atlantic Forest encompasses Brazil's most densely populated areas, including the megacities of São Paulo and Rio de Janeiro. Almost 150 million people, about 70 percent of Brazil's population, live here. On a recent journey from São Paulo to Rio de Janeiro, we drove for hundreds of miles past a relatively uniform, treeless landscape of cattle pastures with degraded and eroding soil; about one cow per hectare scratched a living from the sparse grass growing there. The same area had once been a vibrant, fertile landscape full of life, biodiversity, water, and food. Over the past few hundred years, it has lost most of its fertility, water, and food production capacity. With the proper restoration techniques, however, we can reverse the degradation process. Doing so will provide food for millions and jobs for thousands of people in a booming regenerative agri-food sector and in ecotourism. But how can such a vast area be restored? Where should one even start? More than 30 years ago, the SOS Mata Atlântica initiative was born to turn the fate of the Atlantic Forest around. For the first few years, only a handful of activists paid attention to the issue. Throughout the 1990s, the initiative helped raise awareness about the disappearing ecosystem among hundreds of thousands of people in Brazil, Argentina, and Paraguay.

In April 2009, a larger initiative was formed to bring together all the organizations and scientists behind the restoration efforts. Public communication has shifted from highlighting a crisis to painting a positive vision for a restored landscape across most of the original extent of the Atlantic Forest. The "Pact for the Restoration of the Atlantic Forest" is a collaborative effort of over 350 organizations, including SOS Mata Atlântica, working across 17 biomes along the east coast of Brazil. Collectively, in the past 30 years, they have already restored over 2.5 million acres (1 million ha) of degraded landscapes across the Atlantic Forest, an area equivalent to the size of the city of Hong Kong or nearly 200 times the size of Manhattan. The initiative's goals are even more impressive: restoring the ecology and functionality of the Atlantic Forest across 37.5 million acres (15 million ha), an area the size of the US state of Illinois or larger than the country of Greece.[6] A trinational Secretariat ensures coordination with similar efforts in Argentina and Paraguay.[7] The current pace of restoration compared to the enormous ambition raises obvious questions of timelines. If restoring 1 million ha took 30 years, it would take 450 years to restore 15 million ha at the same pace. We do not have that kind of time to meet the current climate and nature crisis. How do we accelerate the restoration not just incrementally but by orders of magnitude?

We posed this question in a small working group of the advisory board of the UN Decade on Ecosystem Restoration. Together with nongovernmental organizations, think tanks, researchers, and public institutions that are part of the 350 organizations of the Pact for the Restoration of the Atlantic Forest, we first mapped the ecosystem services that could be produced by a restored Atlantic Forest, including clean water, food production, and ecotourism. We then identified the industries and sectors that would benefit from these ecosystem services.

3D Agriculture: An Example from São Paulo

Estimating that restoration would cost approximately 4,000 USD per acre (10,000 USD on average per ha), a restoration investment of around 150 billion USD would be needed across the region to restore 15 million ha. The only feasible funding source at this level would be public–private partnerships. Such partnerships would combine public-sector investments, such as the drinking water restoration efforts in São Paulo discussed in Chapter 4, with private-sector investments, such as the production of agri-food commodities or other industrial goods and services from nature. In a brief report, we listed the key ecosystem services provided by restoration investments and concluded that agroforestry is the most feasible tool for initiating large-scale restoration across the Atlantic Rainforest, because it is relatively fast and low cost, and can produce large amounts of agricultural goods.[8] Then we mapped out a 250,000-acre (100,000-ha) agroforestry pilot project and sought volunteers and investors to bring it to reality. Courageous Land, an aptly named startup, stepped forward as the project lead. The company helps smallholder farmers to shift their plots to regenerative agroforestry, providing them with precise design and planting instructions; for the many landowners who have left the countryside for the city, Courageous Land can plan, organize, and finance land management on their behalf, providing what the organization calls "agroforestry intelligence," a much-needed form of AI. Public-sector financial support could come in the form of blended finance, discussed in Chapter 4, or clear public incentives for private-sector investments into large-scale, economically feasible restoration techniques such as agroforestry.

But what exactly is agroforestry? In one of the pilot plots I visited in January 2024 north of São Paulo, a heavily degraded hillside with eroding and infertile soil is starting to flourish again, after a combination of banana plants, native and exotic timber (later to be replaced by native trees exclusively), coffee, papaya, and over 20 other species were planted. Some call agroforestry "3D agriculture" or "3D forest gardens," in reference to the three overlapping layers of plant life that absorb sunlight and turn it into biomass and agricultural produce.

First is the tree canopy layer, which produces timber and provides partial shade and windbreaks for the next layers. A middle layer of palm trees followed by a lower layer of trees, shrubs, and bushes that can produce coffee, cocoa, berries, and other fruits comes next. The final layer of herbs and vegetables still receives enough sunlight to thrive. Sometimes a fourth layer is added: climbing vines that cover the tree trunks and can produce additional fruits or nuts. A multistory agricultural system like this is more permanent and requires lower recurring investments than annual crops such as wheat. Most plants in agroforestry are perennials, meaning that once they are established, they bear fruit for many years. It is also much more productive in terms of food produced per area than most other forms of monocrop agriculture. Agroforestry is indeed a multisolver solution for the polycrisis we are in. After as little as a few months, agroforestry systems improve soil health, water retention, biodiversity, and productivity.

After we visited the newly established pilot plot, we traveled to a six-year-old agroforestry plot in a neighboring town. Shade trees and banana plants were already higher than the adjacent farmhouse, and the undergrowth was a lush and diverse garden of edible herbs, fruit trees, coffee shrubs, and cash crops such as cassava. Avocado and cocoa trees are planted beneath the initial shade cover of bananas and fast-growing eucalyptus, which the farmers remove and sell as fast-growing timber to make way for a longer-term layer of Indigenous fruit, nut, and native timber trees, further diversifying the growing income of this family farm. We heard over 40 species of birds and saw many kinds of butterflies that had come back to live in this shaded, cool, and moist forest garden; in contrast, in the neighboring cattle pasture, the soil was baked and cracked, and no wildlife was in sight.

The plants in the Courageous Land agroforestry plot include several nut-bearing trees from the Atlantic Forest biome, including the ancient Paraná pine (*Araucaria angustifolia*), an evergreen that has been around for 200 million years and can grow over 160 feet (50 m) tall. These trees will take 15 years to start producing nuts, if they ever do so. Some of the nut-producing trees found in the tropical forests of Brazil cannot be cultivated successfully outside of their natural habitat because the trees require ecological diversity for their reproductive cycle, including symbiotic relationships with orchids and other plants that attract pollinators. Becoming diverse enough to provide the right ecological conditions for the wild nut trees to flower and bear nuts will be a test for this agroforestry system. In the meantime, this forest garden is becoming more productive and more profitable with each passing year. And it produces jobs: At least four workers are employed full time on this small farm of less than 49 acres (20 ha), which generates many more jobs than cattle ranching on degraded pastures. The agroforestry boom around São Paulo and other cities is starting to create a shortage of skilled farm labor, which can be addressed with proper training. Moving toward full employment in Brazil is a welcome long-term trend, as in some rural areas, the unemployment rate is above 10 percent.

The transition from degraded cattle pasture to agroforestry is a fast-start and long-term restoration plan that produces results from the outset. Restoration is a process rather than an end goal, and this piece of land is now under active restoration. It will continue to improve for decades. Over time, it will produce more food, store more carbon in the soil, restore biodiversity above- and belowground, filter and store more water, and provide more livelihoods for members of rural communities. The income option includes carbon credit revenue from selling some of the restoration results. However, as the global market for buying carbon credits is volatile, it is essential to diversify farm income. By mid-2025, more than 550 farms (and growing) had signed up to transform their operations with the help of Courageous Land. But the company's CEO, Phil Kauders, told me he wants to go much further, much faster. "With the help of technology and artificial intelligence, we can map and design agroforestry plots in a fraction of the time it used to take. And we can help farmers to market their produce better and in real time for the best prices," Phil said during our visit to one pilot farm.

Courageous Land is one example of a new industry that did not exist just a few years ago: the nature restoration industry. The company, as well as other Brazilian companies including ReGreen, Mombak, and Biomas, was founded in the past few years and is growing rapidly. In 2025, the federal government began offering restoration concessions on public land, presenting a new and exciting business opportunity for ecopreneurs in Brazil and globally. The level of economic opportunity in Brazil alone is enormous; an estimated 50 million acres (20 million ha) of land is available for

191

World Restoration Flagships

restoration, including large areas of public land. If we assume that perhaps one-third of this land will be restored using agroforestry and that 7 acres (3 ha) of mature agroforestry land can support one full-time job in agriculture, the emerging restoration industry in Brazil could create more than 2 million new jobs, if the government and private sector work together to create the right conditions to enable this ecosystem restoration boom. Governments need to establish the right public incentives and provide clear, long-term land-use rights and legal conditions for investments in restoration, as seen with the recent carbon law in Brazil. That law establishes a national cap-and-trade system that sets greenhouse gas emission limits across multiple sectors, requiring companies exceeding their allocated emissions to purchase carbon credits. The aim is to reduce national emissions and to integrate Brazil's carbon market with international carbon markets. In the law, the Brazilian government recognizes the enormous potential of the "bioeconomy," an economy based on the restoration, conservation, and sustainable use of biodiversity, and is actively creating the right investment conditions.

Achieving World Restoration Flagship Status

Across the Atlantic Rainforest, the more than 350 organizations in the Pacto Mata Atlântica have all the knowledge, experience, and network necessary to restore the vast area, and they have the government's full support. All they need is investment, and all investors need is the design of low-risk, large-scale projects with a reasonable rate of return. With the 250,000-acre (100,000-ha) restoration project we helped to initiate, attracting the required level of investment is now simply a matter of time. Even larger-scale projects are now under development across Brazil. There is more than enough for investors, farmers, civil society, and the government to do to grow this industry. When consumers buy local, seasonal, and diverse produce, they are helping to establish commercially viable regenerative agriculture models. Even in areas that seem too degraded, dry, or eroded, restoration can take hold and spread far and wide. In 2022, the United Nations awarded the Pacto Mata Atlântica the World Restoration Flagship status in recognition of the immense restoration potential across the Atlantic Forest and the impressive results already achieved. Just like the Atlantic Forest, there are countless large landscapes worldwide waiting to be restored to their full health and productivity, if only we could see their potential.

Farther Than the Eye Can See

One of the biggest challenges we face in large-scale landscape restoration is a lack of imagination. When we look at a landscape, we should always question if this is how it has always looked. Is it degraded compared to its potential? How could it become a more productive and thriving landscape for people and wildlife? We should allow ourselves to envision the landscape in a way that enables humans and nature to form a mutually beneficial union. Doing so would make landscapes more beautiful, biodiverse, and valuable to humans, ultimately leading to healthier, happier, and more affluent communities. Let's take the example of the Sinai desert, the landmass connecting Egypt to Asia.

Recent archaeological findings and historic texts have changed our perception that the Sinai landscape has always been dry and barren. This stretch of sparsely populated, arid land is approximately the size of the US state of West Virginia, or slightly larger than the country of Switzerland. Today, the landscape cannot sustain many people or much wildlife. However, once this area was home to a large human population living in a lush and green landscape before deforestation, overgrazing, and changes in water management transformed it into a dust bowl, possibly thousands of years ago.

Regreening the Sinai has recently gathered political and economic momentum. It is one of many global examples of a potential large-scale ecosystem restoration that would change the lives and livelihoods of millions of people and the fate of entire nations. What if we focused our collective human ingenuity, political will, and finances on restoring landscapes at this scale, providing new homes to millions of people, animals, trees, and other wildlife and plants? What if we got started instead of just dreaming about it? It helps to have powerful examples such as the UN World Restoration Flagships to generate the inspiration and required level of ambition.

The World Restoration Flagship Concept

Most of us have no mental image of what ecosystem restoration on a vast scale could look like, how it would be done, or what it might cost. We simply lack recent and well-known examples. We may have a good understanding of significant infrastructure investments, such as major roads, railways, or airports, and it is widely accepted that they are necessary and costly to build and maintain. But what about nature as our critical infrastructure? What if we could repair our forest watersheds to ensure the future drinking water supply? What if we could restore extensive mangrove forests or oyster banks to protect our coastal cities from storms? What about ensuring food security by investing in productive landscapes, healthy soil, stable water tables, and intact forests and wetlands?

The UN Decade on Ecosystem Restoration 2021–2030 allows us to change this gap in our imagination. Our teams at the UN Environment Programme and the UN Food and Agriculture Organization brainstormed early in 2021 on how to build the UN Decade into a global movement that would trigger large-scale investments in nature, based on a new understanding of our relationship with the natural world. We identified the need to nominate and promote large-scale "flagship projects" for inspiration: impressive examples of restoration that could show the art of the possible. Thus, the idea of the UN World Restoration Flagships was born. The flagships aim to shift public perception toward the successful, large-scale transformation that would be feasible if we put our collective minds, hearts, and hands to it.

Together with pioneers among UN member states, civil society, and the private sector, the UN Decade is designating huge areas for successful restoration that can inspire the entire world. By mid-2025, the UN had selected 17 projects on land and in the ocean spanning more than 50 countries. In total, the flagships that have been recognized to date have committed themselves to restore an area of 400 million acres (162 million ha) by 2030, an area the size of the US state of Alaska. More projects are in the pipeline, generating billions of US dollars in investments and spurring a race to the top between countries and regions to invest in restoration.[9] The large-scale ambition is spreading.

In June 2024, for example, after more than a year of fierce political debate, the European Union decided that all EU countries must restore at least 20 percent of their land and sea areas to good condition by 2030. The decision had seemed unthinkable just a few years earlier. The move toward restoration at scale will also trigger an entirely new industry: companies that specialize in restoring thousands of acres of degraded land. As mentioned, the Pacto Mata Atlântica was awarded World Restoration Flagship status in 2022, and Acción Andina, discussed in Chapter 5, in 2023. More examples are waiting to be discovered by the global public, policymakers, and investors. Even in the Amazon forest, despite problems with recent fires, the rate of deforestation is slowing, and restoration efforts are gaining momentum.

The Arc of Restoration in the Brazilian Amazon

The rate of deforestation in the Brazilian Amazon has dropped sharply in two years under the administration of President Luiz Inácio Lula da Silva and is now at its third lowest level in the past two decades.[10] This news is only partly good because we are still losing precious primary tropical rainforest, and we should bring that rate to zero. However, alongside the slowing pace of deforestation, another positive trend is gaining momentum: the large-scale restoration of forests in the heart of the Amazon biome. In December 2023, Brazilian climate scientist Carlos Nobre, Brazil's National Development Bank, and other partners launched an ambitious plan: the Arc of Restoration. This effort focuses on restoring degraded and deforested areas along the southern and eastern edge of the Amazon, a region historically known as the Arc of Deforestation due to intense forest loss. The restoration needs in the Amazon are immense, covering up to 60 million acres (24 million ha) across the Brazilian Amazon and up to 500 million acres (200 million ha) across the entire Amazon basin, which spans nine countries: Brazil, Peru, Bolivia, Ecuador, Colombia, Venezuela, Guyana, Suriname, and French Guiana. In a 2024 publication, the Science Panel for the Amazon outlines strategies for achieving this compelling vision of large-scale Amazonian restoration for Brazil and details the immense benefits such restoration would mean for the region, the country, and the world. Restoration at this scale would stabilize the entire biome from its imminent ecological tipping point.[11]

Fortunately, courageous investors and "ecopreneurs" are rising to the massive challenge of restoring the Amazon and other forests of Brazil. When I met Peter Fernandez, the charismatic CEO of the restoration startup company Mombak, he shared his compelling vision for Brazil to become a global superpower in the carbon markets through large-scale restoration. And true to his word, he is one of the visionary investors and ecopreneurs making this vision a reality. Peter comes from the technology industry, where he served as CEO of 99, Brazil's first unicorn—a startup valued at over 1 billion USD. He cofounded Mombak in 2021 with Gabriel Silva. Since its founding, Mombak has attracted over 300 million USD in investments to restore more than 240 acres (100,000 ha) of the Amazon. Mombak's most significant project to date is a 50,000-acre (20,000-ha) reforestation area in the heart of the Amazon, in the state of Pará. Mombak's projects, and other companies that have recently started restoration of the Amazon and other biomes in Brazil, are still not reaching the scale envisioned by the Arc of Restoration, or the overall commitment of the Brazilian government to restore at least 30 million acres (12 million) ha of degraded forest land, but it is a good start that will inspire other companies and other investors to join the restoration economy.[12]

The cost of restoring the Amazon at a large scale would be high, probably more than 100 billion USD. That sounds like a lot, but it's not much when compared to the cost of inaction. Losing even just 10 percent of its agricultural output would cost Brazil 14 billion USD each year, not counting similar losses in agriculture in neighboring countries.[13] Let's return to the concept of the flying rivers discussed in Chapter 3. Providing rainwater for South America is one of the many benefits of the Amazon rainforest basin. But the biotic pump of the Amazon is at risk due to deforestation. If the flying rivers stopped bringing rain across the continent, agricultural output across Latin America would be severely limited. Maintaining this essential ecosystem service is worth billions of USD per year, in addition to priceless global benefits of avoided climate tipping points. The production value of agriculture in South America accounts for almost 400 billion

USD per year.[14] To maintain the ecosystem, we need to connect the goods and services nature provides with an economic reinvestment in nature, as we saw in Chapter 4.

We could maintain the flying rivers and fund the protection of the Amazon through a levy or other form of recurring payment from the commercial agriculture sector in Brazil and its neighbors. Finding the funding for restoration at scale is all about making the economic value of nature more visible and finding out who benefits. The World Bank and other groups currently are developing ideas for large-scale payments for ecosystem services. Identifying the economic benefits of the Amazon and other large forest biomes as a global good is one of the arguments behind President Lula's Tropical Forests Forever Facility, which would pay countries with extensive intact forests for maintaining them in perpetuity. This facility would probably pay for itself within less than a decade, given agriculture's dependence on rain from forests. At the same time, in the agricultural sector, massive ecosystem restoration projects are also underway, although mostly out of the global public's view. For example, a transformation has begun in southeastern India in recent years that will change our understanding of agriculture as we know it.

Community-Based Natural Farming in India

Located in the southeastern region of India, Andhra Pradesh is one of India's 28 states, with a population of 55 million people. Six million of its inhabitants are classified as "smallholder" farmers, which is usually defined as farmers with less than 12.5 acres (5 ha) of land under management. Together, the farmers of Andhra Pradesh manage 20 million acres (8 million ha), an area about the size of the US state of South Carolina or the country of Austria. Thirty years ago, the government created women-led cooperatives to transition the management of land to more nature-based, regenerative practices while simultaneously increasing individual farmers' incomes and reducing their costs for industrial chemicals and fertilizers.

This effort received a significant boost and broad public support during the second half of the 1990s and the early 2000s, when a wave of farmer suicides took place. Caught in a vicious cycle of decreasing yields, farmers spent more and more money on industrial fertilizers and pesticides, which sent them into a downward spiral of debt. As many farmers saw no economic future for their families, they took their own lives. Between 1995 and 2014, a total of 296,438 Indian farmers died by suicide across India, often from drinking the industrial pesticides that were the cause of their debt.[15] The government rightly saw the current agrochemical industry model of agriculture as a dead end.

The chief minister of Andhra Pradesh, Nara Chandrababu Naidu, has invested heavily over the past 10 years, both in his second term from 2014 to 2019 and since his reelection to his third term in office in 2024, to improve the livelihoods of smallholder farmers while also making agriculture more resilient, more biodiverse, and more self-sufficient by sourcing all ingredients for fertilizers and pest management locally, from native plants like the neem tree. The state government has also stepped in as an aggregator and negotiator of smallholder produce between farmers and large buyers, including the international agri-food giants Olam, Nestlé, and Unilever. Individual farmers would face difficulties negotiating with large buyers, and, in turn, these large companies would need to manage thousands of individual supplier relationships. Through government intervention, both sides win. Everyone benefits from a fair price and a large, long-term, predictable, transparent supply chain.

Over 1 million smallholder farmers of Andhra Pradesh are already engaged in the world's largest agroecology transformation. The state government, civil society, and businesses are collaborating to reach over 6 million farmers in the coming years, aiming to shift them to community-based natural farming methods that increase yields, resilience, biodiversity, and farmer income. The productivity of the entire landscape has increased due to the increased number of trees across the landscape, for fruits, nuts, and timber, as fodder for livestock, or just for windbreaks. New technology, including artificial intelligence, can accelerate this transformation, as AI agents can help provide advice and facilitate meaningful two-way communication with thousands of farmers simultaneously, in multiple languages. The relatively small number of highly skilled technical advisors in Andhra Pradesh has been a limiting factor for the government's agriculture extension service to date, but with new technology, the transformation can spread across the entire state. Other states in India and other countries across the Global South are starting to replicate the success of Andhra Pradesh. There are an estimated 1 billion smallholder farmers in the world. Making them champions of climate and nature action could open new economic opportunities across much of the developing world.

Now let's look at an example in South Africa. The next project is one of my favorite examples showcasing the vision and impact of landscape-level ecosystem restoration. It is situated in and around the picturesque Baviaanskloof Valley, a UNESCO World Heritage Site in the Eastern Cape province of South Africa.[16]

A New Restoration Economy in South Africa

Driving through the Eastern Cape of South Africa, one might mistake the vast, open landscape for savannah; in fact, it is often a degraded version of what was once a dense thicket ecosystem. In this ecological context, one can imagine a thicket as a miniature forest, reaching a height of about 10 to 13 feet (3–4 m) in arid areas; in wetter areas, it can grow to 19 to 22 feet (6–7 m). Although these thickets are relatively short, they are extremely dense—often more densely packed than taller tropical or boreal forests—with shrubs, small trees, vines, and creepers creating an impenetrable landscape.[17]

Today, much of the Eastern Cape is a wide-open and mostly brown landscape sparsely vegetated with old, single trees every few hundred steps for hundreds of miles. One might expect to see a lion or another of the Big Five safari animals on the horizon in the glimmering heat. However, elephants, buffalo, rhinos, lions, and leopards disappeared from much of this landscape long ago. Although the landscape looks picturesque and rugged, it is a heavily degraded ecosystem. The entire Eastern Cape region was once one of the world's best habitats for black rhinos and other Big Five species. However, ecosystem degradation started about 200 years ago, with the widespread introduction of goats and Merino sheep. After two centuries of overgrazing, it is difficult to imagine what this landscape once looked like—and what it could become again.

However, thanks to a concerted government intervention, the carbon market, and payments for water as an ecosystem service, farmers across this 5,020-square-mile (1.3 million ha) area now have the option to restore the degraded land with native forests while maintaining a decent living. The areas under restoration are so large that they are visible from space as bright green dots across the brown, drought-stricken landscape. I had a chance to visit the area in 2019 and speak with some of the pioneers of landscape restoration, including farmer Pieter Kruger in the Bavianskloof Valley.

The End of the Road

Pieter had to make one of the toughest decisions of his life when the goats on his farm had nothing more to eat, the soil had eroded, and most of the vegetation had been destroyed. "I have always been a farmer," he told me, "but that moment in 2007, I knew that I could not go on. There was no more water. Zandvlakte is the last farm in our valley in the Bavianskloof, and the river had run dry before it reached my farm." Pieter reluctantly gave up goat farming and joined the Working for Water programme, a government pilot initiative aimed at restoring degraded watersheds.

Over the next three years, he and a team of over 100 workers planted 3,750 acres (1,500 ha) of his farm with millions of cuttings of spekboom, an indigenous succulent tree that can thrive even in dry conditions (see Figures 7.1 and 7.2). "I have never regretted that decision," Pieter told me in 2019. "The trees are now well established, and in the big flood this year, we managed to keep water runoff from penetrating the soil instead of washing away our topsoil into the river."

Figure 7.1 The "wonder plant" spekboom, a South African native succulent tree that can grow in semiarid landscapes and establishes a dense thicket ecosystem.
Credit: Florian Fussstetter/UN Environment Programme

Figure 7.2   A restored spekboom plot in the foreground of a large landscape waiting to be restored.
Credit: Florian Fussstetter/UN Environment Programme

Spekboom (*Portulacaria afra*) is native to the Eastern Cape region of South Africa. The trees can grow even on steep slopes in mountainous areas. They suck up the moisture quickly and can store it for months, forming living "water tanks" for times of drought. Spekboom forests can serve as grazing and browsing areas of last resort for wildlife and livestock, even when everything else has withered.

Goats or Restoration?

Spekboom, which absorbs carbon dioxide from the atmosphere faster than most other trees in dry conditions, occurs widely across the 12.5 million acres (5 million ha) of the thicket biome and is the dominant plant species in approximately 3.2 million acres (1.3 million ha) of South Africa, an area almost the size of the US state of Hawaii or roughly 60 percent the size of Wales in the United Kingdom. The spekboom plant makes this vast degraded landscape a low-hanging fruit for restoration at large scale and fast speed. "Spekboom is an amazing plant. It can take root and regrow simply from cuttings taken from existing trees. It can quickly reform the soil because it continuously sheds a lot of leaves, which help to build up soil organic carbon," explains ecologist Anthony Mills, who has published extensively on the thicket ecosystem of South Africa, one of the country's lesser-known plant biomes. Spekboom is the keystone plant in the thicket ecosystem: it provides food and shelter for a wide range of other species. Its role goes beyond being a silent climate hero. "Spekboom doesn't just sequester carbon; it transforms harsh brown landscapes into vibrant green ones," Anthony adds. The availability of carbon markets is making this transformation possible. Since 2010, Anthony's company, AfriCarbon, has planted 31 million spekboom cuttings across the Eastern Cape.

The South African government views thicket restoration as one of the most immediate and cost-effective options for achieving national climate and biodiversity goals and recognizes that private investments are key to success. In 2007, it started the most extensive ecological experiment in the world with a goal of restoring an area of thicket of over 2.5 million acres (1 million ha), almost 200 times the size of the US city of Manhattan, providing work and income for thousands of people for several years. "We have studied this thoroughly, and we believe there are big opportunities for ecosystem restoration investments across South Africa," Christo Marais, chief director at the Department of Environmental Affairs, which runs the program, told me at the time.

One of the next steps in scaling up restoration could be establishing a national carbon market—where governments, businesses, or individuals purchase carbon credits in exchange for growing plants and trees—and livestock farms, where several thousand hectares can be replanted with spekboom, and farmers can combine income from carbon credits with other income streams. "Farmers like to look over the fence and see what their neighbor is doing," says Pieter Kruger. "Having big demonstration plots on existing farms is important to spread the word that becoming a carbon farmer can pay off, both for restoring the land and for making a decent return from the land."

A few years after my visit, ecosystem restoration funded by carbon credits became a booming industry across the Eastern Cape, despite carbon market volatility. Stabilizing global and national demand for high-quality carbon credits would provide this massive restoration project with another boost to reach all its benefits in this semiarid environment, including drinking water supply for cities in and around the Eastern Cape, such as Bishho and Gqeberha (Port Elizabeth). The restoration of the entire landscape would secure the nation's long-term future. However, restoration is essential not only in dry conditions. Even in the wettest habitats, such as peatlands, wetlands, and coastal forests called mangroves that are submerged in saltwater for most of their lives, ecosystem restoration is gaining speed and scale.

Building with Nature in Indonesia

Mangroves comprise only 1 percent of the world's forests but can store up to five times as much carbon per hectare as forests on land. They also provide spawning and nursery grounds for about one-third of commercial fish species in the tropics and subtropics. No mangroves, no more fish on the plates of most of the world's population—it's as simple as that. It is not surprising, therefore, that a large fan club is mobilizing around mangrove conservation and restoration, including the team at Salesforce. About 20 million of the 60 million trees we have funded are mangroves.

In 2021, my colleagues at Salesforce and I helped to establish a collective mangrove initiative of over 100 governments, NGOs, and businesses. Led by The Nature Conservancy, Conservation International, and over 30 other NGOs and scientific institutions and backed by over 100 countries and companies, the Global Mangrove Breakthrough aims to generate a 4 billion USD collective effort to conserve and restore 37 million acres (15 million ha) of mangroves, an area about the size of the US state of Florida. This effort would not only help to stabilize the global climate, but it would also save one of the most important ecosystems for the world's fish stocks.[18] Mangrove restoration projects are now being undertaken all around the tropics. In one of the most promising examples, the World Restoration Flagship Building with Nature uses assisted natural regeneration to allow mangroves to rebound naturally in Demak, a low-lying coastal community in Java, Indonesia. The project has reduced erosion, flooding, and land loss caused by subsidence, and this successful approach is expanding along much of Java's coastline. Once tried and tested, this approach could be replicated along most of Indonesia's low-lying coastlines.

Restoring Planet Earth

Other iconic World Restoration Flagships include the Great Green Wall, a restoration corridor across the Sahel region of Africa; the Living Indus project to restore Pakistan's primary water source; the restoration of Mediterranean Forest Landscapes; and a project to help the world's small island developing states such as Vanuatu or Samoa to build sustainable "blue" economies around healthy marine ecosystems. We need many more such flagships in all countries. The ideas and seeds for them exist everywhere. We may not yet be seeing the full potential of a restored Earth. However, with the right inspiration and imagination for planetary-scale restoration, and with a global collective effort, we can halt and reverse the loss of nature. The returns on our initial investment would be enormous, and benefits would accumulate over many generations through four key returns: natural capital, social capital, financial capital, and inspirational capital. Benefits would flow largely to those who need them most: vulnerable communities in the Global South, particularly the 1 billion smallholder farmers on whom food security for much of the world depends.

My vision for this decade, and the decades beyond, is that we will have the courage and wisdom to follow the inspiration of the World Restoration Flagships and scale up the many large and small restoration efforts that are taking place across the world to rebuild a sense of community, a sense of local pride, and our sense of belonging to nature as our common home. The final chapter of this book summarizes my main call to action: an invitation to view nature with different eyes and with an open mind and open heart, as the beginning of a new relationship between humanity and nature that can transform both. The chapter also includes my family's journey toward this new relationship through our ongoing farming adventure. Although we do not control any area the size of a World Restoration Flagship, we are restoring our small farm and our sense of self and purpose through it.

Chapter 8

Stubborn Optimists

*The more powerful, positive and detailed our vision of the
future, the more likely we are to achieve it.*

—Brahma Kumaris, a worldwide spiritual
movement led by women[1]

One of the essential ways we can help a world that seems to
be becoming more dystopian, less free, and more stressful get
back on track is by understanding that where we focus our thoughts
and our attention, we also focus our energy. We need energy to create or achieve anything. If we focus our attention on fear, anxiety,
or possible future loss, we drain ourselves of the energy we need to
deal with crises.

It is easy to get distracted by the negative news that seems to
fill most headlines and to pour our energy into the pits of fear and
anxiety. It is even easier to overlook all the positive developments
happening around the world because often they are less loud, less
aggressive, and less in the news. Yet these positive developments are
everywhere. We need to bundle good news and positive thoughts in
our minds and use them to create a strong, positive vision that we
want to move toward as individuals and as a human society and civilization. A positive vision that can keep us motivated, focused, and
energetic. And we need those more than ever in times of crisis.

Throughout human history, positive guiding frameworks traditionally have been provided by storytelling or religion (and more

recently written fiction). However, positive, modern visions of living in harmony with nature, like *Laudato Si'*, discussed in Chapter 3, are rare in world religions. Many faiths are losing followers to the secular, consumption-focused, and hectic modern lifestyle. And fiction is not playing its role of providing humanity with a clear vision right now because the main tool of contemporary fiction, the novel, primarily deals with moral dilemmas of individuals, not all of society. The modern novel is ill equipped to paint a comprehensive vision for all of humanity. Admittedly, we can draw conclusions for humanity as a whole from the unlikely heroes in the few recent novels that address nature loss and climate change. Consider *The Overstory* by Richard Powers or *Flight Behavior* by Barbara Kingsolver, who both feature lead characters who are struggling to cope with the overwhelming new reality of climate chaos and nature loss.[2] However, we cannot solve the crisis we face collectively at the individual level. In his important book *The Great Derangement*, Amitav Ghosh outlines how today's modern novel is so focused on exploring our individual choices and moral dilemmas that it cannot provide us with a compelling vision for the future of society as a whole.[3] As far as a new narrative for humanity is concerned, it seems we are on our own for the time being. No individual hero or story will emerge to save us from ecological collapse. We all have to be heroes now, just like at any other time in human history, and move beyond our focus on individual self-actualization to a renewed sense of collective action and community. This final chapter discusses how to be our own heroes and write our own story of a future of peace and prosperity rooted in a new relationship with nature.

The Future We Want

The United Nations attempted to create a blueprint for humanity's future with its 2015 adoption of the Sustainable Development Goals, aptly calling it *The Future We Want*. However, instead of a compelling vision, it is a technical plan consisting of 17 goals with 169 detailed targets, covering everything from ending world hunger to decent work and clean energy.[4] The effort is laudable and useful, but without a compelling narrative to accompany the plan, it is like having the blueprint for building a ship on which we could all sail into a beautiful future together when no one has told the story of the destination that inspires people to build a ship, change their lives, and set sail. Until there emerge new and powerfully positive narratives of an inclusive and peaceful future for humans and nature alike, each of us will have to paint that vision for ourselves and compare and share it with others. Let's start with the most essential ingredient for positive change: a positive mindset.

Optimism Is Mandatory

I would like you to meet Pando, a tree living in the mountains of the state of Utah in the United States. Pando—which is Latin for "I spread"—is, in fact, more than just a tree. It is a forest of about 47,000 aspen trees, all originating from and living off the same root system. They are one single being of genetically identical trees.[5] The entire being of Pando weighs approximately 6,000 metric tons and spans 106 acres (43 ha). New research from the Georgia Institute of Technology and the University of Chicago just revealed that Pando might be between 16,000 and 80,000 years old.[6] If so, Pando is possibly the oldest and one of the largest living beings on our planet. Pando not only survived countless forest fires, storms, and likely some close meteorite hits, but it might also have survived an ice age. At a sufficiently large scale, and with the right management, ecosystems are incredibly resilient. Whenever I feel the need for a dose of optimism, I think of Pando and the numerous other examples of extreme resilience in nature.

In building a new and stronger relationship with nature, optimism is one of the first and most essential lessons we can learn. Nature makes me optimistic because the sheer will to survive and thrive puts all bad news into perspective. Another valuable lesson from Pando and all other natural processes is humility. This ancient and vast organism survived severe disturbances over millennia, is self-replicating, scrubs the atmosphere of excess carbon dioxide for free, and produces only clean air, clean water, and new fertile soil in the process—besides being a great tourist destination. It is clearly smarter and better designed than any technology that we humans have ever built.

With some luck, Pando will live for thousands of years more. Currently, its main threat is the fact that deer and livestock consume too many of the young saplings. Fortunately, that is a problem that could be solved. A new management plan for Fishlake Basin, Pando's home, foresees fencing, cattle guards, and road decommissioning to improve tree regeneration.[7] The reintroduction of large predators does not seem to be an option for Fishlake National Forest because of conflicts with livestock grazing. However, in other areas where overgrazing by deer is also an issue for forest health, the introduction of wolves or other large predators could solve the problem. Wolves act as a keystone species, meaning while they are not the most numerous animal in an ecosystem, their presence affects and changes the composition and structure of entire landscapes by shifting the behavior and number of their prey.

For nature's ingenuity and resilience, you don't necessarily need to look for ancient forests near you. Just observe how powerfully and quickly nature bounces back from disturbance and even from degradation. With some practice, if we watch and listen carefully, we can still perceive nature's vibrant energy and diversity across time, even in heavily degraded landscapes. A recent restoration technique in the United Kingdom, for example, involves identifying glacial meltwater ponds, or *pingos*, that are now often covered by several soil layers. When researchers uncovered the ponds and filled them again with water, rare freshwater plant seeds germinated after more than 10,000 years.[8] Nature is patient and resilient, and nature remembers. You can be part of reviving nature, building on the memory still stored in ecosystems. Nature is never static but always evolves and changes, and we can be active agents of positive change. Visit a conservation or restoration site near you, spend some time volunteering, and be inspired by nature's optimism. Nature is always optimistic because optimism is the only way to actively create a better future. It is the only mindset that allows for evolution and growth.

Taking action is the most effective way to maintain hope and optimism. Action, hope, and optimism fuel and reinforce each other. However, there are two kinds of optimism, according to economist Paul Romer. He explains: "Complacent optimism is the feeling of a child waiting for presents."[9] It depends on external good news. That is not the kind of optimism we need right now. Even when good news is hard to come by, we can remain optimistic as a matter of choice. The kind of optimism we need is called conditional optimism, "the feeling of a child who is thinking about building a tree-house. 'If I get some wood and nails and persuade some other kids to help do the work, we can end up with something really cool.'" We all need to be conditional optimists now. Let us strive to cultivate this mindset and remind ourselves daily that every thought and word is a form of energy that makes a difference in the world.

More Bloom, Less Gloom

The environmental movement, after decades of partial victories and many lost battles, has almost forgotten how to win. A pervasive sense of doom, rather than a focus on creating a better world, limits our potential to shape the future we want. Yet we can relearn to be conditional optimists. We can change the world, and we will. In fact, we are doing so all the time, with every thought, word, and action. The question is just in which direction we are changing it. What is the world we want to live in? To me, it is a world where our relationship with nature has been healed, and we can live in natural abundance. How can we get there? "If we could change ourselves, the tendencies in the world would also change," Gandhi said. Yet it seems to me that for every 100 people I meet who want to change the world, only one person wants to change themselves first. I am not saying we can save the world through individual choices alone. Rather, significant, rapid collective change occurs when enough of us are changing the way we think, feel, and perceive the world. The world as it is, or as we perceive it, flows out of the state of our hearts and minds. If we change our hearts and minds, we will certainly change the world. You can try this in small steps: One day, keep an optimistic attitude toward everything and everyone, keep smiling, and be cheerful. Then, on another day, try to be grumpy, negative, and scowl at everything and everyone in your path. Compare the results at the end of both days. It is a universal law that we shape the world by how we perceive and interpret it.

Go Slow to Move Fast

At several points throughout this book, I have conveyed a sense of urgency for us to take faster and more decisive collective action. And although that is true for society at large, and for some of the necessary changes in our economy and our sectors of industry, another layer to the crisis we are in requires the exact opposite: slowing down. At an individual level, we are caught up in a world that is becoming faster, more hectic, and more distracting. I met a young monk from the Brahma Kumaris spiritual movement at the UN Climate Change Summit in Paris in 2015 who told me, "To stop global warming, we all just have to chill." I have the feeling that this is exactly the advice nature would give us if we asked the question and listened carefully for an answer. Nature knows that the calmer and more grounded we are, the more powerful is our ability to withstand and recover from disturbance. Finding a strong core of inner peace, regardless of external circumstances, is a lesson from nature that seems counterintuitive to a crisis situation, which cries for fast and urgent action. However, we can bring our best to this situation individually only if we first stop and breathe. I meditate each morning. Even though sometimes it is difficult to find the time or the concentration, it is a necessary preparation for my busy days. Meditating makes my days more productive, more creative, and more fun. It is also a reminder that there is more to human existence than the material world. There is a spiritual realm to which we are always connected, even though we might not know or feel it. If we treat nature with respect and approach nature with a slow, open, and calm mindset, we can be reminded of this bond and our connection to the infinite. And, sometimes, a crisis can be useful to remind us of the need to seek inner balance.

A Journey Forward to Nature

For many people, including myself and my family, the COVID-19 crisis threw into sharp relief what is truly important in life. We spent the initial COVID period at our home in Nairobi, Kenya. As lockdowns alternated with periods of restricted travel, and the UN Environment Programme office remained closed for an indefinite period, we relocated to my wife's home country, Denmark. We wanted to be near our aging parents in that country and in Germany and, at the same time, give our two children the chance to experience life in Europe, beyond the usual brief holidays we spent there each summer. While in Denmark, we found ourselves in a fortunate situation that allowed us to realize a long-held dream. Ever since my wife and I met, now

Stubborn Optimists

well over 20 years ago, we had dreamed of buying a farm and living and working close to nature. Our international lifestyle, with stints in Brussels, Montreal, and Nairobi, had prevented us from searching for and finding our piece of land. With the COVID shift to remote working, we suddenly had the time to search for the right place and make it our own. In December 2020, we bought an old dairy farm, with 90 acres (35 ha) of cropland and some forests. It is located close to our families and within easy reach of a small international airport in central Denmark that would keep us connected to the world and a train station in our small village with a direct connection to Copenhagen and other cities.

Relocating from the bustling city of Nairobi, with its near-perfect climate and international flair, to a tiny village in Denmark, where it is often cold and wet, was not an easy transition. However, as we began to see the first signs of progress in restoring our piece of land and finding purpose in sharing our regenerative farming experience, we have become increasingly grateful for the chance to reset our relationship with nature. Over the past few years, we have planted a diverse forest, dug out two lakes, established hedgerows and a fruit tree alley, and converted our fields from cropland to pasture, using a rotational grazing system with sheep. In the future, we plan to move to a silvo-pastoral system with fruit and nut trees and grazing of sheep, ducks, and chickens. Our operation is still small, and we are feeling our way forward toward commercially viable regenerative farming. We realize that this transition is possible only because we can subsidize it with other income, which, by the way, is true for most farming in the European Union: Many farmers have second jobs or can make ends meet only due to government subsidies. A way is needed for farmers to invest in the transition to regenerative farming without raising significant funds on their own. Part of this transition would support the establishment of local markets where farmers can sell directly to consumers (current EU rules make such direct sales difficult) and get a fair price for their produce. Regenerative farming can be more profitable for farmers because it relies much less on expensive external inputs, such as fossil fuel–based fertilizers and industrial pesticides. However, shifting from agriculture based on chemistry to agriculture based on biology takes some time; soils that are essentially dead have to be revived first. And regenerative techniques like permaculture require different market structures, because they tend to produce small quantities of many different products throughout the year. Permaculture is a form of regenerative agriculture that employs a combination of tree species, rotational livestock grazing, and organic agriculture to regenerate and sustain soil biodiversity and enhance the health and productivity of agricultural systems. If financing can be found for this evolution in modern agriculture and the required market structures, entire landscapes would be transformed and nature would be restored on a massive scale. On our small farm, we can already see how quickly nature responds to the intentionality we bring to restoring the land and the soil.

Putting Culture Back into Agriculture

Our farm is not only a way to restore our relationship with nature. It is also an exploration of our culture and our sense of local community. The word "agriculture" originates from the combination of two Latin terms: *ager*, meaning "field" or "land," and *cultura*, meaning "cultivation" or "growing." In many parts of the world, much of human culture it rooted in the annual agricultural cycle, with festivities and religious celebrations revolving around the harvest, planting, and birthing cycles. Agriculture is only one of several economic sectors that currently extract value, health, and resources from natural systems and humans at a breakneck speed, without replenishing them, with detrimental consequences for both present and future generations.

Our ability to interact directly with nature—such as knowing how, when, and where to sow, plant, and grow our food or how to prepare it—has been largely outsourced to a sprawling industrial agri-food system, with significant impacts on the ecological balance of local ecosystems worldwide. At the same time, the hyperindustrial food chain we rely on has severely limited the diversity of our food, making our global food supply less resilient, less nutritious, and less local than it could be. An estimated 30,000 edible plants exist, and approximately 7,000 species have been cultivated for food. Yet most of our current food system relies on just 30 species of plants, and 40 percent of all the calories we consume in the world today come from just three staple crops: wheat, maize, and rice.[10] This enormous simplification for the purpose of industrialization of our most fundamental human cultural good is concerning. The highly industrialized food system arguably has made food cheaper today than it has been at most other times in history. Yet it has impoverished our diets, limited nutrition, and made our global food system much more fragile. And although food is relatively cheap today, people also choose to spend less of their income on food. For example, in the United Kingdom, the proportion of household spending on food halved from 33 percent in 1957 to 16 percent by 2017.[11] At the same time, our agri-food system with its growing proportion of highly processed foods is causing other costs and negative consequences, such as malnutrition and obesity (which both are on the rise, ironically); soil erosion; animal cruelty; pollution of soils, water, and air; groundwater depletion; climate destabilization; and nature's demise.

These massive costs to society are not reflected in the real price of food, as we saw in Chapter 4. The UN Secretary-General was clear in his statement to the UN Food Systems Summit in 2021: "Global food systems are broken—and billions of people are paying the price."[12] Agriculture and food systems are responsible for over a quarter of global greenhouse gas emissions and more than half of biodiversity loss.[13] Toxic runoff from our predominant form of energy- and input-heavy industrial agriculture now even threatens life in the ocean by accelerating coral bleaching and harmful algae blooms. What can we do?

Fortunately, there is a way to produce food that is healthier for humans and for the rest of nature. An increasing number of remarkable turnarounds from debt-ridden conventional farms to productive and profitable regenerative farms worldwide are occurring. These farms always have one thing in common: They focus on healthy soils.

The Soil Food Web

No amount of industrial chemical input can replace the natural functioning of soil biodiversity. Natural allies of farming, such as earthworms and microorganisms, keep the soil porous, allowing water to infiltrate and be stored, and recycle and unlock essential nutrients. Instead of building on this natural cycle, industrial agriculture relies heavily on external energy, resulting in massive soil erosion, rapid loss of soil fertility, and the runoff of nitrogen and other nutrients into bodies of water. Lakes and rivers become choked with fertilizers, causing algae blooms and the loss of freshwater biodiversity.

Healthy soil is a cornerstone of any healthy terrestrial ecosystem, influencing everything from crop yields and food security to carbon sequestration and overall biodiversity. However, intensive agriculture, and related excessive fertilizer use, chemicals, and intensive plowing, have led to a severe decline in soil biology, and thereby also in soil fertility and productivity, worldwide. Physical disturbances, such as plowing and other forms of tilling the soil, as well as pesticide use are the primary problems that disrupt soil structure and kill the soil microbiome. Plowing is also a major cause of soil erosion. The UN Food and Agriculture Organization estimates we have as little as 50 years of topsoil left.[14] Fortunately, new science and technology exist that can allow us to regenerate soil health and shift to no-till or low-till agriculture.

Agriculture done well is a blend of natural processes and human ingenuity that benefits our health and biodiversity and helps to stabilize the climate. Recent studies of regenerative agriculture in Europe, for example, have shown that, compared to conventional farms, permaculture farms across Central Europe have 27 percent more soil organic carbon—the component in soil with the capacity to store carbon that originally comes from atmospheric CO_2.[15] Soil organic carbon enables plants to grow and soil biodiversity to thrive. In the study, the soil was also more porous, essential for rainwater infiltration, and had more earthworms—201 percent more. The concentrations of various vital nutrients were higher on permaculture sites, indicating better crop production conditions. And the diversity of birds was 197 percent higher on permaculture sites. We can have healthy nature and eat it too!

The key to regenerating soil lies in understanding and restoring the soil food web. The term is derived from the food chain. Most organisms eat (and are eaten by) more than just one other organism, meaning we all rely on a complex, multidimensional food web rather than a linear food chain. Such a food web also exists under our feet. Often referred to as nature's operating system or the soil biome, the complex web of life below the surface comprises living organisms from insects and earthworms to microscopic fungi, bacteria, protozoa, and nematodes. These organisms interact with each other and with plants, performing crucial ecological functions, including unlocking nutrients for plant growth. A balanced soil food web acts as nature's nutrient cycling system, breaking down organic matter and soil particles to release nutrients in a plant-available form, allowing plants to control the flow of nutrients they need. Most soils are derived from a variety of parent materials, such as rocks, sand, silt, and clay, which contain atoms of essential nutrients, such as iron, boron, phosphorus, calcium, potassium, and trace elements, that plants require. An active soil microbiome helps plants access nutrients from these crystalline mineral structures. The soil biome pulls nitrogen, another essential element for plant growth, into the ground from the air, which is 78 percent composed of nitrogen. Many plants, such as clover, alfalfa, peas, and groundnuts, have symbiotic relationships with nitrogen-fixing bacteria. A healthy soil microbiome also provides natural protection against pests, diseases, and weeds. It reduces the need for irrigation and builds soil structure that prevents erosion and compaction. Plants growing in healthy soil are much healthier themselves and require fewer or no pesticides.

Modern intensive agricultural practices are disrupting the soil food web and compromise its natural functions, leading to depleted soil, increased reliance on chemical inputs, declining yields, and environmental damage. Restoring the soil food web solves these problems, enabling farmers to operate without chemicals, dramatically increasing yields, reducing costs, improving crop resilience, sequestering carbon, and protecting waterways and biodiversity. The more life exists in the soil, the more soil organic carbon is drawn from the atmosphere into the soil, and the more food the soil can produce. Just the upper 3 feet (1 m) of soil contains about 50 percent more carbon than the entire atmosphere, and the amount of organic carbon in all forms of soil could be increased significantly. The 4 in 1,000 initiative launched by the French government at the Paris Climate Summit estimates that we would need to increase soil organic carbon only by 0.4 percent per year globally to absorb all fossil fuel–related emissions.[16]

Boosting soil biodiversity can make this possible. Dr. Elaine Ingham is a leading soil biologist who has pioneered research in this field. Her Soil Food Web Approach is a trademarked, three-step method to regenerate soil biology rapidly. (For more information,

Stubborn Optimists

see: www.soilfoodweb.com.) After Soil Food Web experts identify which groups of microorganisms are lacking in a sample, farmers can add tailor-made composts that contain the missing microorganisms. The last step is the adoption of natural farming techniques for soil biology survival.

The focus on the power of soil to save the future of farming, with saving nature and stopping global warming as welcome side benefits, is now starting to spread. In his fabulous four-part documentary *Roots So Deep*, Peter Byck from Arizona State University shows the struggle of conventional farmers with debt, nature loss, and an uncertain economic future.[17] Byck has led a 10-year, multidisciplinary science project to explore how cattle ranching can turn from a culprit of climate change to a solution. Adaptive multi-paddock (AMP) grazing is a technique that mimics nature's process of building soils and maintaining soil health. Rotating a large number of livestock across grasslands resembles the natural rotation of huge herds of large herbivores across open landscapes. Instead of overgrazing and degradation, which can result from long-term grazing on the same paddock, frequent rotations of the herds after periods of intensive grazing stimulate the perennial grasses to grow deeper roots and store more carbon in the soil. Compared to conventional grazing, AMP grazing builds up higher amounts of soil organic carbon while also improving soil biodiversity and the bird and insect diversity that depends on healthy soil. Best of all, farmers no longer need to buy expensive nitrogen-based or other fertilizers because the soil microbiome, in combination with cattle dung and urine, unlocks all the nutrients the plants need exactly when they need them.

In the documentary, Byck and the team from Arizona State University and 12 other science institutions across the United States and the United Kingdom compare five conventional cattle farms in the southeastern United States with five neighboring farms that used

AMP grazing. Across the five farm pairs, the AMP side produced 13 percent more soil carbon and 9 percent more soil nitrogen stocks (without synthetic fertilizers), three times more grassland birds, more than double the rainwater infiltration, 33 percent more diverse and numerous insects, and 25 percent more numerous and active microbes.[18] All these outcomes contribute to a healthier and more profitable farm operation than their conventional neighbors. It is a win-win-win.

Well, almost. There is one loser, and it is a powerful and sore loser. Big agricultural-chemical (agri-chem) companies stand to lose a multi-billion-dollar market if the profits from agriculture shift back to the farmers rather than to the companies. No wonder that Bayer and other agri-chem companies are trying to claim the term "regenerative agriculture" for themselves. Beware of greenwashing, which is starting to be prolific in this space. If there is a big fertilizer, chemical, and fossil fuel bill to pay at the end of the transition process, it is not regenerative farming. Some of these companies are realizing that they need to completely reinvent themselves. The fact that beef might be carbon neutral in future, or perhaps even climate and nature positive, is an exciting prospect for many agri-food companies.

There is, of course, a role for big business in the future of agriculture. It is just a different role from the current industry that produces millions of tons of toxic pesticides and fertilizers. New technologies, such as AI-powered soil acoustic detectors, measure soil health by the sounds made by the tiny denizens of the underworld. Invertebrates, such as worms and beetles, make sounds within the topsoil that can be used to gauge the biological activity, providing fast insights without the need for soil lab analysis. Robots and precision agriculture tools and software can take on weeding, mulching, and planting that are currently done mostly by manual labor. Although these novel technologies are still rather expensive, at least for our small farm, prices are coming down fast. Someday we will be able to measure the life and vitality in an ecosystem as easily as we measure the flow of electricity through a circuit board. The future is already here.

More Than Dirt

The massive die-off of countless soil microorganisms and other farm biodiversity worldwide over the past few decades has caused significant stagnation and loss of biological productivity in the soil. Healthy soils can contain more than 1 billion microorganisms in a teaspoon of soil; in one cup, there would be more microorganisms than people on Earth.[19] The good news is that it is not hard to get these trillions of unseen workers back on every farm. We simply need to let nature be our ally in agriculture rather than our enemy. In her book *Wilding*, Isabella Tree describes the transformation of the Knepp Estate in Sussex, UK, from a debt-ridden, destructive industrial farming complex into a commercially viable, internationally admired "rewilding haven."[20]

By allowing nature to take over many essential farm processes, such as soil regeneration, water retention, and drainage, the farm saved millions of pounds each year. Once the massive flywheel of a functioning, diverse ecosystem based on healthy soil starts to turn at a sufficient scale and speed, it shifts into an equilibrium that allows for the careful harvesting of surplus in perpetuity, with the added benefits of maintaining biodiversity, clean water, and more nutritious produce. We began this transition on our farm in 2022, and it is already yielding results.

We recently took several soil samples from various parts of the farm and had them analyzed by an expert from the Soil Food Web, who found that our soil was already bouncing back. The analyst was surprised that no microorganisms whatsoever could be found in a soil sample from an industrially managed field that our neighbors own. Without biological activity to loosen and continuously turn over the soil, it quickly becomes a hard crust during a drought that can absorb and store significantly less rainwater. The "helpers" in our soil work around the clock to unlock nutrients, recycle carbon, absorb and store rainwater, and increase soil resilience and productivity. Conventional farmers have to replace all these free ecosystem services with expensive external inputs, ranging from industrial fertilizers to deep tilling of the soil with heavy machinery and a significant use of fossil fuels. Despite this heavy investment of financial capital, most conventional farms lose significant natural capital each year, in the form of soil erosion, soil compaction, and groundwater depletion.

Nature Returns to Our Farm

In just a few years, nature is bouncing back on our farm. The first year we owned the farm, we established new hedgerows as well as a strip of flowering plants alongside our fields. The insect diversity that emerged that year was astounding; there even were some rare insects, such as hawk moths, whose flight and feeding habits resemble those of hummingbirds. None of our neighbors had ever seen one before. Hawkmoths probably had been absent from our area for years, but they came back as soon as they were invited. Nature is patient, resilient, and bounces back quickly if given the space. Imagine if all households and farms were paid to introduce native flowering plants. We could quickly bring back insect diversity in most places, including the bees to pollinate California's almond trees, at much lower cost than the current 500 million USD per year rent-a-bee model discussed in Chapter 4.

In the second year, we planted a forest on about 20 acres (8 ha) that used to be an intensively managed monocrop maize field, largely devoid of wildlife. It is now teeming with flowering plants, birds, and insects that thrive among the 30,000 seedlings of 26 native tree and shrub species we have planted. When designing the forest, we opted for distinct clusters of a few tree species instead of mixing all 26 species across the entire area. This approach will increase the area's recreational value for our local community, as the shape and composition of the forest will change during a walk. For example, we planted an "Edible Forest" featuring wild cherries, wild apples, walnuts, hazelnuts, and chestnuts. The "Little Sweden" part of the forest looks like a northern Scandinavian forest with mostly birch and pine trees. And we have dense stands of evergreen, coniferous spruce, and pine where deer and birds can seek refuge in winter. And we also ignored conventional forestry advice that, before planting trees, we should spray the whole field with the pesticide Roundup, to kill all "competing" weeds, and then deep plow the field. Instead, we sowed a cover crop of flowering plants to keep the grass at bay and planted the trees directly into the unplowed soil. Doing this probably saved the trees' lives, because right after planting, we had almost eight weeks without proper rain. Plowing would have destroyed the capillary system that kept some moisture in the upper soil. And if we had killed all other plant life, the bare soil would have dried up much more quickly. Our trees survived because we followed nature's processes.

With some government support, we also dug out two minor lakes on our fields. The larger one is about 0.6 acres (2,500 m²) and was filled with water only three months after we completed the digging. The other one is still a bit more of a sandpit rather than a pond, but we remain hopeful. And even if it will not completely fill with water, it will be a good habitat for amphibians and insects. When we monitored bat echolocation calls just a few months after we dug out the new lake and wetland, 8 of the 15 species of bats native to Denmark were already present and circling the water at night to catch insects.

On the fields that we converted from maize monocultures to grassland and are managing through rotational grazing by our sheep, we have observed significant improvements in soil fertility and in biodiversity both above and below the soil. The number of swallows, larks, and other insect-eating birds has almost doubled in three years. I was particularly pleased this year when, four years after I had hung up an owl nesting box inside one of our barns, we spotted a pair of beautiful barn owls. We will continue to shift our farm to a permaculture system that does not rely on industrial fertilizers or pesticides. Our aim is to become a productive, resilient farm that produces high-quality food and inspiration for ourselves and others. Our vision for our land and landscape is to restore bio-diversity, do our part in stabilizing the climate, and promote local resilience in our community and happiness within our family. Our farm might be small, but it has trillions of workers. Only a few of them are human. The rest are the unseen nematodes, fungi, soil bac-teria, and the birds, insects, and other animals that keep the living ecosystem functioning. To succeed in our effort to restore nature, both at our small scale and for the planet as a whole, we just have to commit to a deeper, trust-based relationship with nature, and in this case with the soil, and allow the flywheel of life to run again at full capacity. As in any relationship, a restored relationship with nature requires trust and respect first and foremost.

Generation Restoration from Local to Global

A new relationship with nature is possible everywhere, in every com-munity and every walk of life. You don't need to live on a farm or in a forest. Nature is always with us, because we are part of nature. And we can heal nature, just as nature can heal us. If we start to see nature with new eyes, nature will respond differently to us. We humans are not just some of the most numerous large mammals on the planet; we are more than just spectators of nature and her processes. We are ecosystem engineers, just like beavers, elephants, or oysters. We have the power to build and shape entire ecosystems, and many other species depend on us and count on us. But unlike any other ecosystem engineer on the planet, our species now has an almost limitless power to create or destroy.

It is possible to shift the view of ourselves from a species that is mostly good at destroying, depleting, or degrading nature to one that is mostly good at nurturing, building, and shaping nature into a lush and abundant garden for all life on Earth. Even those who believe that our current market economy can exist only if we continue to squeeze more and more efficiency out of nature should be convinced by this new viewpoint, because a new relationship with nature does not come at the cost of abundance and a prosperous lifestyle. In fact, it is the only way to make abundance and prosperity possible for humanity in the long run. If we restored nature to an order of magnitude more productivity, diversity, and abundance than we have today, vastly more economic value would be one of the many beneficial results.

Once we realize that we are part of nature and acknowledge our role and responsibility as humans, we will have started our new relationship. This change in mindset will spark the human curiosity, creativity, and collaboration that is necessary to restore Planet Earth. With the energy and action that our new relationship with nature will unlock, we can build a legacy that lasts for generations: a greener, cleaner, safer, and more just world for all. Let us be #GenerationRestoration. There is much to share, to learn, and to gain for everyone.

Epilogue: The Freedom to Choose

We live in an exciting time in human existence. At the dawn of an artificial intelligence revolution that will forever change the way we live and work, our future is simultaneously threatened by a triple planetary crisis of climate change, nature loss, and pollution, a crisis that is an outward expression of a more profound crisis of the human spirit. Our global civilization, with all its shortcomings, has also allowed billions of people to live a life in dignity and relative economic prosperity and safety. The past 80 years since World War II, often called the Long Peace, have seen a dramatic reduction in wars between major powers and a general decline in violent deaths as a proportion of the global population. However, without a long-term, compelling, and clear vision of what it means to be human and how we can live a life of dignity and peace with all other life-forms on Earth, we could be slipping back into the survival mode and competition over resources that has characterized much of human history. The geopolitical situation is fragile, and our ability to develop and deploy technologies, including weapons of mass destruction, is unparalleled in history. The fact that we are, for the first time, living as a single global social group, with the means to communicate in real time across all of humanity, has torn down many social boundaries and certainties. A backlash against this overwhelming new reality has driven many social groups, mostly nation-states, into a retreat of "us versus them" mindset. The thin fabric of social cohesion that holds our global civilization together is tearing.

However, there is also immense hope and opportunity. Our ability to connect all of humanity through technology allows a global discussion about the future we want. The rise of artificial intelligence reminds us to ask the fundamental questions that have always been central to human existence: Who are we? Who do we want to be? What is our purpose and role in life and on this Earth? AI can perhaps write better books, tell better stories, sing better songs, and soon do many things better or at least more efficiently than humans can. However, AI does not yet possess consciousness, and we can utilize this historical window to evolve our understanding of what it means to be human. I am still searching for many of the answers myself, but of one thing I am certain: We must make peace with nature for *Homo sapiens* to survive and thrive. It is clearly one of the necessary next steps in human evolution.

Throughout this book, I have shared personal stories of my search for meaning and the human connection with nature. I would also like to conclude this book on a personal note. One of the most fascinating aspects of life is that every human being can have a unique relationship with nature. I am sharing my perspective on the spiritual aspects of nature here, because it is one of the reasons why I wrote this book. I am not asking or expecting anyone to share my views about nature or spirituality. To me, life does not make much sense without the concept of a benevolent intelligence that is creating order in the cosmos. Since the singularity that created our universe, a fantastic level of structure, order, and harmony has emerged through the natural laws of physics. With our emerging knowledge of quantum mechanics, science can now start to explain things that were long confined to the realm of faith: that we are both energy and matter, and that our energetic state is connected with other energies across space and time. I believe the amazing level of sophistication of our world is evidence enough that there is a loving intelligence or intelligent love that structures the universe for us, around us, and, most surprisingly of all, together with us. We can call this loving intelligence by many different names. Most world religions discourage us from turning our idea of this loving intelligence into a specific image or concept, because it is too powerful for the human mind to grasp, and any description would not do it justice.

Nature, including our bodies and all of our material world, is an expression of that loving intelligence. We have been given the abilities and the free will to create our own expressions and stories on this cosmic canvas, with the power to create and destroy. Throughout history, our fate as individuals and as human societies has always been in the balance between these two forces. We walk on that thin edge between heaven on Earth or hell on Earth, and although we sometimes veer more one way than another, both individually and collectively, we never go all in for one side. Because that is not the point of human existence. The point is that we are free to choose a life in peace, harmony, and justice over a life in greed, anger, and confrontation. Sometimes that choice is not fully within our control as individuals when society pushes us one way or another. However, we always have a choice to influence which way we go.

By acknowledging and realizing that we are part of nature, we are choosing the path of peace, of love, and of life. We are ready to play our rightful role within creation, to the joy of one another and to the joy of the cosmic force that we are all connected to. Our assigned role is that of responsible stewards of creation, and when it comes to nature, that means we should act as deliberate and benign ecosystem engineers. We are the guardians and gardeners of nature. Aligning with the intention of all of creation to increase harmony, balance, and beauty in this world, we can play our part in the drama of life on Planet Earth by restoring nature as much, and as widely, as possible.

The next time you reflect, pray, meditate, or worship, think about nature as part of creation and yourself as part of nature. Some of us are deeply linked to a spiritual path, whatever it may be, and some of us are not. Most of us still have a long way to go toward becoming better human beings. Making peace with nature is a necessary and beautiful part of that journey for everyone. Caring for nature is simply the next great step in human evolution. As Jane Goodall encouraged us in her foreword: "Over the millennia, Mother Earth has nurtured us, and now she needs our help. Let us move forward with passion, determination, and hope and work together to heal and care for Planet Earth. Let us enter into a new era of moral and spiritual evolution."

Let us take that step now and heal our relationship with nature and with each other. Nature is waiting. It is time to come home.

Taking Action

Most nonfiction books on climate change or nature include long to-do lists. This book does not, because its central message is that, first and foremost, we need to think and feel differently about nature and our place within it before we act. The urge to act and appropriate actions will naturally flow from a new mindset that humans are part of nature. I welcome and encourage anyone, anywhere, to stand up for the conservation and restoration of nature, no matter their religion, age, ethnicity, or sexual or political orientation. Nature is home for all of us and has room for all of us. Here is a list of a few excellent initiatives, in alphabetical order, to take practical action and deepen our relationship with nature.

ActNow: The United Nations platform to suggest and track sustainable actions that we all can take in our daily lives. www.un.org/en/actnow

Ecosia: The search engine that uses ad revenue to plant trees worldwide. www.ecosia.org/

Ecosystem Restoration Communities: A network of local communities restoring earth for a sustainable future. www .ecosystemrestorationcommunities.org/

Good On You: Rates fashion brands on sustainability, labor rights, and animal welfare. Helps users shop for clothes that match their values. https://goodonyou.eco/

JouleBug: Gamifies sustainable habits with challenges and social sharing. www.joulebug.com/

Patagonia Action Works: Find action-oriented environmental initiatives near you, and connect with grassroots campaigns and volunteering opportunities. www.patagonia.com/actionworks/

Plant for the Planet: A donation platform and youth ambassador network for global forest conservation and restoration. www.plant-for-the-planet.org/

Project Drawdown: A leading research organization that identifies, evaluates, and shares the most effective climate solutions available today, helping individuals, businesses, and policymakers take measurable action to reach drawdown: the point when greenhouse gases in the atmosphere begin to decline. https://drawdown.org/

ReGeneration: A global initiative and online platform that mobilizes people, organizations, and communities to take collective climate action and accelerate solutions for environmental and social regeneration. https://regeneration.org/

Restor: The platform Restor.eco maps the largest number of active conservation and restoration sites across the globe. You can set up a profile and upload your own initiative, no matter the size or location, and learn about restoration efforts near you. www.restor.eco

Roots and Shoots: Jane Goodall's youth movement and network offer engagement opportunities for people of all ages to take action for people, animals, and the environment through local projects and campaigns. https://rootsandshoots.global/

Seafood Watch: Find seafood that's fished or farmed in environmentally sustainable ways. www.seafoodwatch.org/

Soil Food Web: Offers online courses and resources on soil biology and regenerative soil management, based on Dr. Elaine Ingham's pioneering work on soil health. www.soilfoodweb.com/

UN Carbon Offset Platform: Companies, organizations, or individuals can purchase units (carbon credits) to compensate for greenhouse gas emissions or simply to support action on climate. https://unfccc.int/climate-action/united-nations-carbon-offset-platform

UN Decade on Ecosystem Restoration: The global movement and digital hub coordinated and backed by the United Nations to prevent, halt, and reverse ecosystem degradation worldwide. The platform provides resources, best practices, and opportunities to join restoration initiatives. www.decadeonrestoration.org/

Wealthsimple and Betterment: Investment platforms offering recommendations and tools to invest in sustainable companies and portfolios aligned with environmental, social, and governance (ESG) criteria.

www.wealthsimple.com/en-ca/managed-investing/socially-responsible-portfolio

www.betterment.com/socially-responsible-investing

We Don't Have Time: A social network to review the climate performance of companies, join campaigns, and connect with global changemakers to drive environmental accountability and solutions. https://app.wedonthavetime.org/

WWOOF (World Wide Opportunities on Organic Farms): Connects volunteers with organic farms around the world, facilitating hands-on learning about sustainable agriculture, food systems, and ecological living in exchange for room and board. https://wwoof.net/

And because saving the world should be fun, I add my favorite family movies and documentaries on our relationship with nature and with the land and sea:

The Biggest Little Farm **(2018).** This documentary, directed by John Chester, follows a couple transforming depleted land into a thriving, biodiverse farm, illustrating the challenges and triumphs of regenerative agriculture. It was part of the inspiration for our own farm.

Common Ground **(2024) and** *Kiss the Ground* **(2020).** These star-filled documentaries, directed by Josh Tickell and Rebecca Tickell, discuss the role of healthy soil in mitigating climate change, highlighting regenerative agriculture as a solution to restore ecosystems.

Fantastic Fungi **(2019).** This documentary, directed by Louie Schwartzberg, explores the hidden world of fungi, their role in soil health, nutrient cycling, and ecosystem restoration.

The Nature of Farming **(***Landbrugets Natur***, 2025).** This film, directed by Isabelle Denaro, showcases Danish farmers using regenerative methods to rebuild soil fertility, biodiversity, and ecosystem function. It makes farming cool again.

Ocean **(2025).** David Attenborough's masterpiece details the urgent need to reverse the degradation of the ocean, the largest part of our blue planet. It was directed by Toby Nowlan, Keith Scholey, and Colin Butfield.

Princess Mononoke **(1997).** This Japanese animated fiction movie, directed by Hayao Miyazaki, explores the conflict between industrialization and the natural world, with a strong message about the need for harmony and restoration.

***Roots So Deep, You Can See the Devil Down There* (2024).** This documentary, directed by Peter Byck, highlights the results of a 10-year research journey into regenerative cattle ranching across the southeastern United States.

***To Which We Belong* (2021).** This documentary, directed by Pamela Tanner Boll and Lindsay Richardson, profiles farmers and ranchers using regenerative practices to restore soil and landscapes, emphasizing hope and practical solutions.

Notes

Chapter 1: Squandering Our Natural Wealth

1. Safina, C. (2015). *Beyond Words: What Animals Think and Feel.* Holt.
2. Fuller, B. (2020). *Operating Manual for Spaceship Earth.* Lars Müller Publishers [1969].
3. National Ocean Service, "Ocean Facts," 2024. oceanservice.noaa.gov/facts/.
4. Quoted in Roberts, C. (2010). *The Unnatural History of the Sea.* Island Press.
5. Erma Hermens, *Crossing and Turning: The Sea Turtle Trade in the 17th-Century* [blog]. https://lookingthroughartblog.wordpress.com/2020/06/03/crossing-and-turning-the-sea-turtle-trade-in-the-17th-century/. Retrieved December 27, 2024.
6. This is my own conservative estimate based on the average oyster reef density of 300 to 500 oysters per square meter, described by Roger L. Mann, Melissa Southworth, Juliana M. Harding, and James A. Wesson, "Population Studies of the Native Eastern Oyster, Cassostrea Virginica (Gmelin, 1791) in the James River, Virginia, USA. Department of Fisheries Science, Virginia. scholarworks.wm.edu/vimsarticles/405/.
7. (2009). The billion oyster project. *Journal of Shellfish Research* 28 (2): 193–193. www.billionoysterproject.org/harbor-history.
8. Kulansky, M. (2006). *The Big Oyster: History on the Half Shell*, 20. Ballantine Books.
9. McCann, M. (2018). *New York City Oyster Monitoring Report: 2016–2017.* The Nature Conservancy, May 2018.

10. Sustainable Eel Group, "Quantifying the illegal trade in European glass eels (*Anguilla anguilla*): Evidences and Indicators," SEG-Report:2018-1-V1, January 2018. www.sustainableeelgroup.org/wp-content/uploads/2018/05/SEG-Report-2018-1-V2.pdf.

11. World Wildlife Fund, "Press Pause on Deep Seabed Mining," https://wwf.panda.org/discover/our_focus/oceans_practice/no_deep_seabed_mining/. Retrieved June 2, 2025.

12. Deppe, H.-J. (1985). "Entenkojen und Entenzug—Versuch einer Auswertung der Fangergebnisse nordfriesischer Entenkojen" [Duck decoys and duck migration—an attempt to evaluate the catch results of North Frisian duck decoys]. *Die Vogelwelt* 106: 1–24.

13. Government of Schleswig-Holstein, *Störe in Schleswig-Holstein: Vergangenheit—Gegenwart—Zukunft* [Sturgeon in Schleswig-Holstein: Past—present—future]. November 2014. www.schleswig-holstein.de/mm/downloads/LLUR/Stoerbuch_online.pdf.

14. McClenachan, L. (2009). Documenting loss of large trophy fish from the Florida keys with historical photographs. *Conservation Biology* 23 (3): 636–643. https://doi.org/10.1111/j.1523-1739.2008.01152.x.

15. Lovejoy, T.E. and Nobre, C. (2018). "Amazon tipping point" [editorial],. *Science Advances* 4 (2): https://doi.org/10.1126/sciadv.aat2340.

16. New York City Department of Environmental Protection, "New York Harbor Survey Program: Celebrating 100 Years 1909–2009." City of New York, 2009. www.nyc.gov/html/records/pdf/govpub/6118new_york_harbor_survey_program_celebrating_100_years.pdf.

17. Deutloff, J., Held, H., and Lenton, T.M. (2025). High probability of triggering climate tipping points under current policies modestly amplified by Amazon dieback and permafrost thaw. *Earth System Dynamics* 16 (2025): 565–583. https://doi.org/10.5194/esd-16-565-2025.

18. Douglas P. Swain and Ghislain A. Chouinard, *Viability of the Southern Gulf of St. Lawrence Cod Population*, Research Document 2008/018. Department of Fisheries and Oceans Canada, Gulf Fisheries Centre, 2008. https://waves-vagues.dfo-mpo.gc.ca/library-bibliotheque/334905.pdf.

19. Henley, B.J., McGregor, H.V., King, A.D., and Linsley, B.K. (2024). Highest ocean heat in four centuries places great barrier reef in danger. *Nature* 632: 320–326. https://doi.org/10.1038/s41586-024-07672-x.

20. UNESCO, "Great Barrier Reef: Australia to Put in Place Urgent Safeguarding Measures Requested by UNESCO" [press release], June 6, 2023. whc.unesco.org/en/news/2570.

21. Smits, D.D. (1994). The frontier army and the destruction of the buffalo: 1865–1883. *Western Historical Quarterly* 25 (3): 312–338. www.jstor.org/stable/971110.

22. Sources for material regarding the Dust Bowl include PBS Learning Media, *A Man-Made Ecological Disaster*. www.pbslearningmedia.org/resource/ecological-disaster-ken-burns-dust-bowl/ken-burns-the-dust-bowl/. Retrieved December 27, 2024. Also Wikipedia contributors, "Dust Bowl," *Wikipedia, The Free Encyclopedia*. https://en.wikipedia.org/wiki/Dust_Bowl. Retrieved December 27, 2024.

23. Long, J. and Siu, H.E. (2018). Refugees from dust and shrinking land: tracking the dust bowl migrants. *Journal of Economic History* 78 (4): 1001–1033. https://doi.org/10.1017/S0022050718000591.

24. Ken Burns, *Black Sunday: The Dust Bowl* [film]. www.pbs.org/wgbh/americanexperience/features/dust-bowl-black-sunday/. Retrieved December 28, 2024.

25. Government of Iowa, *Iowa State Economic Profile 2023*. www.ibisworld.com/united-states/economic-profiles/iowa. Retrieved December 28, 2024.

26. Abraham Parrish, *Climate Migrants of the 1930's Dust Bowl* [Library of Congress Blogs], December 1, 2023. https://blogs.loc.gov/maps/2023/12/climate-migrants-of-the-1930s-dust-bowl/.

27. Howlader, A. (2023). Determinants and consequences of large-scale tree plantation projects: evidence from the Great Plains Shelterbelt project. *Land Use Policy* 132: https://doi.org/10.1016/j.landusepol.2023.106785.

28. Sears, R. and Perrin, W.F. (2018). Blue Whale: *Balaenoptera musculus*. In: *Encyclopedia of Marine Mammals* (ed. B. Würsig, J.G.M. Thewissenm, and K.M. Kovacs), 43–43. Academic Press.

Notes

29. International Whaling Commission, *Blue Whale (Balaenoptera musculus)*. https://iwc.int/blue-whale. Retrieved December 28, 2024.

30. The UN Framework Convention on Climate Change covers seven main greenhouse gases; CO_2 is only one of them. The others are methane (CH_4), nitrous oxide (N_2O), hydrofluorocarbons (HFCs), perfluorocarbons (PFCs), and sulphur hexafluoride (SF_6), and nitrogen trifluoride (NF_3). In order to have one common "currency" of estimating emissions, and because CO_2 is the most important greenhouse gas in terms of volume and impact, the unit of CO_2 equivalent, or CO_2e, is commonly used. Pearson, H.C., Savoca, M.S., Costa, D.P., and Roman, J. (2022). Whales in the carbon cycle: can recovery remove carbon dioxide? *Trends in Ecology and Evolution* 38: 238–249. pubmed.ncbi.nlm.nih.gov/36528413/.

31. Bar-On, Y.M., Phillips, R., and Milo, R. (2018). The biomass distribution on earth. *Proceedings of National Academy of Sciences* 115 (25): 6506–6511. https://doi.org/10.1073/pnas.1711842115.

32. UN Environment Programme, *Inclusive Wealth Report 2023: Measuring Sustainability and Equity*, September 23, 2023. https://wedocs.unep.org/20.500.11822/43131.

33. UN Environment Programme, *Becoming #GenerationRestoration: Ecosystem Restoration for People, Nature and Climate*, June 3, 2021. https://wedocs.unep.org/bitstream/handle/20.500.11822/36251/ERPNC.pdf.

34. J. B. Mackinnon, "A 10 Percent World," *The Walrus*, September 12, 2010. https://thewalrus.ca/a-10-percent-world/.

35. IPBES (2019). *Global Assessment Report on Biodiversity and Ecosystem Services of the Intergovernmental Science-Policy Platform on Biodiversity and Ecosystem Services* (ed. E. Brondizio, S. Diaz, J. Settele, and H.T. Ngo). Bonn, Germany: IPBES Secretariat https://doi.org/10.5281/zenodo.3831673.

36. FAO, *Global Forest Resources Assessment 2020–Key Findings*, 2020. Rome. doi.org/10.4060/ca8753en.

37. Sophie Ledger, Claire Anna Rutherford, Charlotte Benham,. . .and Anna Staneva, *Wildlife Comeback in Europe: Opportunities and Challenges for Species Recovery. Final Report to Rewilding Europe by the Zoological*

Society of London, *BirdLife International and the European Bird Census Council*, September 2022. 10.13140/RG.2.2.24283.44324.

38. Langhammer, P.F., Bull, J.W., Bicknell, J.E., and Brooks, T.M. (2024). The positive impact of conservation action. *Science* 384 (6694): 453–458. https://doi.org/10.1126/science.adj6598.

39. Hallmann, C.A., Sorg, M., Jongejans, E., and de Kroon, H. (2017). More than 75 percent decline over 27 years in total flying INSECT biomass in protected areas. *PLoS ONE* 12 (10): https://doi.org/10.1371/journal.pone. 0185809.

40. BirdLife International, *State of the World's Birds 2022: Insights and Solutions for the Biodiversity Crisis*, September 27, 2022. www .birdlife.org/papers-reports/state-of-the-worlds-birds-2022/.

Chapter 2: A Century of Ecology

1. For full disclosure: I joined the board of the Global Evergreening Alliance in September 2024.

2. Richardson, K., Steffen, W., Lucht, W. et al. (2023). Earth beyond six of nine planetary boundaries. *Science Advances* 9 (37): https://doi.org/10 .1126/sciadv.adh2458.

3. Simard, S. (2021). *Finding the Mother Tree: Discovering the Wisdom of the Forest*. Knopf.

4. Lovelock, J.E. and Margulis, L. (1974). Atmospheric homeostasis by and for the biosphere: the Gaia hypothesis. *Tellus* 26 (1–2): 2–10. https:// doi.org/10.1111/j.2153-3490.1974.tb01946.x.

5. The Anthropocene is not yet recognized by the International Commission on Stratigraphy as an official Earth epoch. The commission is the primary body responsible for defining and naming Earth's geological time units.

6. Personally, I am skeptical of some elements of the Gaia theory, but for the purpose of illustrating a few basic ecological principles, please bear with me and imagine for a few minutes that you are Planet Earth. For more on the Gaia theory, see: Wikipedia contributors, "Gaia Hypothesis," *Wikipedia, The Free Encyclopedia*. https://en.wikipedia.org/w/index.php?title= Gaia_hypothesis&oldid=1293508371.

7. IPCC (2023). *Climate Change 2023: Synthesis Report. Contribution of Working Groups I, II and III to the Sixth Assessment Report of the Intergovernmental Panel on Climate Change* (ed. Core Writing Team, H. Lee, and J. Romero). Geneva, Switzerland: IPCC https://doi.org/10.59327/IPCC/AR6-9789291691647.

8. Naughten, K.A., Holland, P.R., and De Rydt, J. (2023). Unavoidable future increase in West Antarctic ice-shelf melting over the twenty-first century. *Nature Climate Change* 13: 1222–1228. www.nature.com/articles/s41558-023-01818-x.

9. Such as the Medieval Warm Period from about 950 to 1250 CE, which saw major expansions of Scandinavian populations, including the Viking settlement of Greenland and North America.

10. Miller, K.G., Raymo, M.E., Browning, J.V. et al. (2019). Peak sea level during the warm Pliocene: errors, limitations, and constraints. *PAGES Magazine* 27 (1): 4–5. https://doi.org/10.22498/pages.27.1.4doi.org/10.22498/pages.27.1.4.

11. von Goethe, J.W. (1819). *West-Östlicher Divan*. Cotta. In the original German the poem reads: "Wer nicht von dreitausend Jahren sich weiß Rechenschaft zu geben, bleibt im Dunkeln unerfahren, mag von Tag zu Tage leben.".

12. Perlin, J. (2022). *A Forest Journey: The Role of Trees in the Fate of Civilization*, 3rd rev.e. Patagonia Books.

13. Hursthouse, R. and Pettigrove, G. Virtue Ethics. In: *The Stanford Encyclopedia of Philosophy*, Fall 2023e (ed. E.N. Zalta and U. Nodelman). https://plato.stanford.edu/archives/fall2023/entries/ethics-virtue/.

14. Plato (1943). *Plato's Republic*. Books, Inc.

15. Shin, S., van Riper, C.J., Stedman, R.C., and Suski, C.D. (2022). The value of eudaimonia for understanding relationships among values and pro-environmental behavior. *Journal of Environmental Psychology* 80: https://doi.org/10.1016/j.jenvp.2022.101778.

16. United Nations, "The United Nations Decade on Ecosystem Restoration: Strategy," 2021. https://globalseaweed.org/wp-content/uploads/2024/12/UN-Decade-on-Ecosystem-Restoration-Strategy.pdf#:~:text=On%

201%20March%202019%2C%20under%20Resolution%2073%2F284%2C%20the,halt%20and%20reverse%20the%20degradation%20of%20ecosystems%20worldwide.

17. Wilson, E.O. (1984). *Biophilia*. Harvard University Press.

18. Nigel Dudley and Sue Stolton, *Running Pure: The Importance of Forest Protected Areas to Drinking Water.* World Bank/WWF Alliance for Forest Conservation and Sustainable Use, August 2003. https://documents1.worldbank.org/curated/en/414701468765561300/pdf/292830Running0pure.pdf.

19. International Energy Agency, "Kenya Energy Mix," n.d. https://www.iea.org/countries/kenya/energy-mix. Retrieved May 31, 2025.

20. A friend of mine, Carlos Calderón Guerrero, wrote his PhD thesis on the impacts of urban trees on air pollution in Madrid, the capital of Spain. He found that each city tree on average filters about 2 pounds (1 kg) of airbound toxic heavy metals and other microscopic harmful pollutants out of the air each year. See Guerrero, *Urban Trees and Atmosphere Pollutants in Big Cities: Effects in Madrid.* PhD thesis, 2014. doi.org/10.20868/UPM.thesis.32872.

21. This program is one of many funded by Salesforce, the global technology company that I joined in May 2022. Sustainability is a core value of Salesforce, and since its origins in 1999, the company has granted almost 1 billion USD to projects (880 million USD as of 2024) that save nature, fund education, and take climate action.

22. American Forests, Woodland Trust, and Centre for Sustainable Health Care, "Tree Equity Score UK," n.d. https://uk.treeequityscore.org/. Retrieved December 17, 2024.

23. Natalie Bicknell Argerious, "In Order to Achieve Tree Equity, the U.S. Must Plant 522 Million Trees in Urban Areas," *The Urbanist*, July 20, 2021. www.theurbanist.org/2021/07/20/in-order-to-achieve-tree-equity-the-u-s-must-plant-522-million-trees-in-urban-areas/.

24. US Forest Service/US Department of Agriculture, Urban and Community Forestry Grants—2023 Grant Awards, 2023. https://www.fs.usda.gov/managing-land/urban-forests.

Notes

25. "Counting trees in the Amazon," *Nature* 502 (2013). https://doi.org/10.1038/502412a.

26. Yana Marull, "Amazon's Flying Water Vapor Rivers Bring Rain to Brazil." https://phys.org/news/2012-09-amazon-vapor-rivers-brazil.htm.

27. Hirota, M., Flores, B.M., Betts, R. et al. (ed.) Resilience of the amazon forest to global changes: assessing the risk of tipping points," Chapter 24 in. In: *Amazon Assessment Report 2021*, C. Nobre, A. Encalada, E. Anderson et al., *Amazon Assessment Report* 2021. New York: United Nations Sustainable Development Solutions Network. Available from: www.theamazonwewant.org/spa-reports/. https://doi.org/10.55161/QPYS9758.

28. Government of Brazil, "August 2024 Amazon Deforestation Lowest in Six Years," September 16, 2024. www.gov.br/secom/en/latest-news/2024/09/august-2024-amazon-deforestation-lowest-in-six-years.

29. Bunyard, P.P. and de Laet, R. (2024). *Cooling Climate Chaos: A Proposal to Cool the Planet within Twenty Years*. BP International.

30. Mo, L., Zohner, C.M., Reich, P.B. et al. (2023). Integrated global assessment of the natural forest carbon potential. *Nature* 624: 92–101. https://doi.org/10.1038/s41586-023-06723-z.

31. Verhoeven, D., Berkhout, E., Sewell, A., and van der Esch, S. (2023). *The Global Cost of International Commitments on Land Restoration*. Wiley.

32. Tom Crowfoot, "Costs for Climate Disasters to Reach $145 Billion in 2025, and Other Nature and Climate News," World Economic Forum, May 7, 2025. www.weforum.org/stories/2025/05/costs-climate-disasters-145-billion-nature-climate-news/.

33. UN Environment Programme, *Nature-Based Solutions for Climate Change Mitigation*. November 4, 2021. Nairobi. wedocs.unep.org/xmlui/bitstream/handle/20.500.11822/37318/NBSCCM.pdfi.

34. UN Environment Programme, *Broken Record: Emissions Gap Report, 2023*. UNEP, Nairobi. https://wedocs.unep.org/bitstream/handle/20.500.11822/43922/EGR2023.pdf?sequence=3&isAllowed=y.

35. UN Environment Programme, *Nature-Based Solutions for Climate Change Mitigation*.

Chapter 3: Nature Is Us: A Tale of Reciprocity

1. Integrative HMP (iHMP), Research Network Consortium (2019). The integrative human microbiome project. *Nature* 569: 641–648. https://doi.org/10.1038/s41586-019-1238-8.

2. For a fascinating and more complete deep dive into this topic, I can recommend Enders, G.'s book (2015). *Gut: The Inside Story of Our Body's Most Underrated Organ.* Greystone Books.

3. Hou, K., Wu, Z.-X., Chen, X.-Y. et al. (2022). Microbiota in health and diseases. *Signal Transduction and Targeted Therapy* 7 (135). https://doi.org/10.1038/s41392-022-00974-4.

4. Altveş, S., Yildiz, H.K., and Vural, H.C. (2020). Interaction of the microbiota with the human body in health and diseases. *Bioscience of Microbiota, Food and Health* 39 (2): 23–32. https://doi.org/10.12938/bmfh.19-023.

5. *Merriam-Webster.com Dictionary*, s.v. "nature." www.merriam-webster.com/dictionary/nature. Retrieved May 31, 2025.

6. *Oxford English Dictionary,* "nature (n.), sense IV.11.a." https://doi.org/10.1093/OED/9458925751. Retrieved June 2025.

7. IPBES, *Summary for Policymakers of the Global Assessment Report on Biodiversity and Ecosystem Services of the Intergovernmental Science-Policy Platform on Biodiversity and Ecosystem Services.* IPBES Secretariat, Bonn, Germany, November 25, 2019. https://doi.org/10.5281/zenodo.3553579.

8. Genesis 1:26–31 (KJV). www.kingjamesbibleonline.org/Genesis-1-26/.

9. Hasna Soussan, "'He Laid Out the Earth for All Living Creatures': Islam's Lessons for Climate," *African Arguments,* April 20, 2023. https://africanarguments.org/2023/04/he-laid-out-the-earth-for-all-living-creatures-islam-lessons-for-climate.

10. Descartes, R. (1996). *Meditations on First Philosophy.* Translated by John Cottingham. Cambridge University Press [1641].

11. Descartes, R. (2006). *Discourse on the Method of Rightly Conducting One's Reason and of Seeking Truth in the Sciences.* Translated by Ian Maclean. Oxford University Press [1637].

12. Hobbes, T. (1996). *Leviathan* (ed. R. Tuck). Cambridge: Cambridge University Press [1651].
13. Spinoza, B. (1996). *Ethics*. Translated by (ed. E. Curley). Penguin [1677].
14. Kant, I. (1997). *Groundwork of the Metaphysics of Morals*. Translated by (ed. M. Gregor). Cambridge University Press [1785].
15. Safina, K. (2015). *Beyond Words: What Animals Think and Feel*. Holt.
16. Thoreau, H.D. (2016). *Walden*. Macmillan Collector's Library [1854].
17. Convention on Biological Diversity, Conference of the Parties to the Convention on Biological Diversity, *15/4. Kunming-Montreal Global Biodiversity Framework*, Montreal, December 7–19, 2022. www.cbd.int/doc/decisions/cop-15/cop-15-dec-04-en.pdf.
18. The 30 by 30 campaign has been very successful in part because it has been well funded: "Philanthropist Hansjörg Wyss Celebrates Historic 30 × 30 Commitment Made by Nations at UN Biodiversity Conference in Montreal," December 19, 2002. www.wyssfoundation.org/news/philanthropist-hansjrg-wyss-celebrates-historic-30x30-commitment-made-by-nations-at-un-biodiversity-conference-in-montreal.
19. Tom Oliver, "A New Campaign Wants to Redefine the Word 'Nature' to Include Humans—Here's Why This Linguistic Argument Matters" [blog, University of Reading], May 10, 2024. https://research.reading.ac.uk/research-blog/2024/05/10/a-new-campaign-wants-to-redefine-the-word-nature-to-include-humans-heres-why-this-linguistic-argument-matters.
20. Francis, *Laudato Si'* [encyclical letter]. The Holy See, May 24, 2015. www.vatican.va/content/francesco/en/encyclicals/documents/papa-francesco_20150524_enciclica-laudato-si.html.
21. António Guterres, "Secretary-General's Address at Columbia University: 'The State of the Planet'," United Nations, December 2, 2020. www.un.org/sg/en/content/sg/statement/2020-12-02/secretary-generals-address-columbia-university-the-state-of-the-planet-scroll-down-for-language-versions.
22. United Nations, *The United Nations Decade on Ecosystem Restoration Strategy*, "Strategy," 2021. UNEP, Nairobi. https://wedocs.unep.org/bitstream/handle/20.500.11822/31813/ERDStrat.pdf?sequence=1&isAllowed=y.

23. Oliver, "A New Campaign Wants to Redefine the Word 'Nature.'"
24. Quoted in Magdaléna Rojo, "Big Oil Wants Nemonte Nenquimo's Ancestors. Not On Her Watch," *Atmos*, October 3, 2024. https://atmos.earth/big-oil-wants-nemonte-nenquimos-ancestors-not-on-her-watch/.
25. Quoted in Astrid Arellano, "'Indigenous Women in the Amazon Must Be Empowered': Interview with Nemonte Nenquimo," *Mongabay*, October 7, 2024. https://news.mongabay.com/2024/10/indigenous-women-in-the-amazon-must-be-empowered-interview-with-nemonte-nenquimo/.
26. Nemonte, N. and Anderson, M. (2024). *We Will Be Jaguars: A Memoir of My People*. Abrams Press.

Chapter 4: The Value of Nature

1. The Danish Climate and Forest Fund, 2025. www.klimaskovfonden.dk.
2. Dasgupta, P. (2021). *The Economics of Biodiversity: The Dasgupta Review*. HM Treasury. assets.publishing.service.gov.uk/media/602e92b2e90e07660f807b47/The_Economics_of_Biodiversity_The_Dasgupta_Review_Full_Report.pdf.
3. UN Department of Economic and Social Affairs (2025). *Handbook on Management and Organization of National Statistical Systems*, 4e, Version 2025/A. United Nations.
4. United Nations, European Union, Food and Agriculture Organization of the United Nations, International Monetary Fund, Organisation for Economic Co-operation and Development, and The World Bank. *System of Environmental Economic Accounting 2012—Central Framework*. United Nations, 2014. https://unstats.un.org/unsd/envaccounting/seeaRev/SEEA_CF_Final_en.pdf.
5. For example, consider the blog from the Institute and Faculty of Actuaries, *Nature-Related Risks Are as Detrimental as Climate Risks: A Call to Action for Actuaries*, October 1, 2024. blog.actuaries.org.uk/nature-related-risks-as-detrimental-as-climate-risks-call-to-action-for-actuaries/.
6. IPBES, *Thematic Assessment Report on the Interlinkages among Biodiversity, Water, Food and Health of the Intergovernmental Science-Policy Platform on Biodiversity and Ecosystem Services*. P. A. Harrison,

Notes

P. D. McElwee, and T. L. van Huysen. IPBES secretariat, Bonn, Germany. https://ict.ipbes.net/ipbes-ict-guide/data-and-knowledge-management/ citations-of-ipbes-assessments/nexus-assessment.

7. Chris Knight, "What Is Natural Capital?" European Investment Bank, June 22, 2023. https://www.eib.org/en/stories/nature-environment-pollution.

8. Wetlands International Indonesia, "Mitigation and Adaptation Roadmap of Land Subsidence in Lowland Coastal Areas in Indonesia," n.d. https:// indonesia.wetlands.org/publication/mitigation-and-adaptation-road-map-of-land-subsidence-in-lowland-coastal-areas-in-indonesia/. Retrieved January 20, 2025.

9. IPBES, *Thematic Assessment of the Interlinkages Among Biodiversity, Water, Food and Health.*

10. UN Environment Programme (2024). *From Grey to Green: Better Data to Finance Nature in Cities. State of Finance for Nature in Cities 2024.* United Nations, Nairobi. https://wedocs.unep.org/handle/20.500.11822/ 46453.

11. International Monetary Fund, "Climate Change and Fossil Fuel Subsidies." 2025. www.imf.org/en/Topics/climate-change/energy-subsidies.

12. UN Nations Environment Programme, *Becoming #GenerationRestoration: Ecosystem Restoration for People, Nature and Climate*, June 3, 2021. https://wedocs.unep.org/bitstream/handle/20.500.11822/36251/ ERPNC.pdf.

13. UN Environment Programme, *State of Finance for Nature: The Big Nature Turnaround—Repurposing $7 Trillion to Combat Nature Loss*, December 2023. https://doi.org/10.59117/20.500.11822/44278.

14. Alberto Méndez Rodríguez, "Experiencia de Costa Rica en el Pago de Servicios Ambientales," Colombia, May 2011. www.ucipfg.com/repositorio/ MAS/CURSO-01/unidad5/MINAET_FONAFIFO%202011.pdf.

15. Dasgupta, *Economics of Biodiversity.*

16. The São Paulo watershed program is run by SABESP (Saneamento Básico do Estado de São Paulo), the largest sanitation company in Brazil and one of the largest in the world. More here: www.sabesp.com.br.

17. The Nature Conservancy, "SABESP (Nature-Based Solutions for Water Management): São Paulo, Latin America & Caribbean," 2024. https://resilientwatershedstoolbox.org/projects/sabesp-nature-based-solutions-water-management.

18. Emily Underwood, "Small Wonders: The Plight and Promise of California's Native Bees," *Flora Magazine*, California Native Plant Society, June 24, 2021. www.cnps.org/flora-magazine/small-wonders-the-plight-and-promise-of-californias-native-bees-23883.

19. Ellen Topitzhofer, Carolyn Breece, Dan Wyns, and Ramesh Sagili, "Revenue Sources for a Commercial Beekeeping Operation in the Pacific Northwest," OSU Extension Service, May 2020. https://extension.oregonstate.edu/catalog/pub/pnw-742-revenue-sources-commercial-beekeeping-operation-pacific-northwest.

20. White, M.P., Alcock, I., Grellier, J. et al. (2019). Spending at least 120 minutes a week in nature is associated with good health and wellbeing. *Science Reports* 9 (7730): https://doi.org/10.1038/s41598-019-44097-3.

21. Jimenez, M.P., DeVille, N.V., Elliott, E.G. et al. (2021). Associations between nature exposure and health: a review of the evidence. *International Journal of Environmental Research and Public Health* 18 (9): 4790. https://pubmed.ncbi.nlm.nih.gov/33946197/.

22. Duncan Murdoch, "US Doctors Are Prescribing Nature in 35 States," Nature Connection Guide, n.d. https://natureconnectionguide.com/us-doctors-are-prescribing-nature-in-34-states/#:~:text=As%20an%20ANFT%20Certified. Retrieved June 30, 2025.

23. VELUX, "Future Generations Face Health Risks from Life Indoors," May 15, 2018. https://press.velux.com/future-generation-of-brits-faces-health-risks-from-life-indoors/.

24. Task Force on Nature-related Financial Disclosure, "Our History." https://tnfd.global/about/history/. Retrieved June 29, 2025.

25. This so-called 4Returns Framework, developed by Commonland, is my favorite narrative for investing in nature. See: https://commonland.com/4-returns-framework/.

Notes

26. Jessica Smith, Diana Denke, Vincent Tang,. . .Romie Goedicke den Hertog, "Private Finance for Nature in 2024: Scaling, Moving Up the Capital Continuum and Connecting to Impact," UN Environment Programme Finance Initiative, Geneva, June 2024. www.unepfi.org/wordpress/wp-content/uploads/2024/06/Nature-finance-overview.pdf.

27. World Economic Forum & McKinsey Company, *Nature Finance and Biodiversity Credits: A Private Sector Roadmap to Finance and Act on Nature*, Insight Report, World Economic Forum, October 2024. www3 .weforum.org/docs/WEF_Nature_Finance_and_Biodiversity_Credits_ 2024.pdf.

28. Willem Ferwerda, "The Trillion-Dollar Promise of a Landscape Restoration Industry," *Forbes*, March 1, 2024. www.forbes.com/sites/forbeseq/ 2024/03/01/the-trillion-dollar-promise-of-a-landscape-restoration-industry/.

29. UN Climate Change, "United Nations Carbon Offset Platform," n.d. https://unfccc.int/climate-action/united-nations-carbon-offset-platform. Retrieved May 31, 2025.

30. Patrick Greenfield, "Revealed: More Than 90% of Rainforest Carbon Offsets by Biggest Certifier Are Worthless, Analysis Shows," *The Guardian*, January 18, 2023. www.theguardian.com/environment/2023/jan/18/ revealed-forest-carbon-offsets-biggest-provider-worthless-verra-aoe.

31. Verra, "Technical Review of West et al. 2020 and 2023, Guizar-Coutiño 2022, and Coverage in Britain's Guardian," January 31, 2023. https:// verra.org/technical-review-of-west-et-al-2020-and-2023-guizar-coutino-2022-and-coverage-in-britains-guardian/.

32. Guy Turner, Jamie Saunders, Utkarsh Akhouri, and Jamie Lambert, *Frozen Carbon Credit Market May Thaw as 2030 Gets Closer*, MSCI [blog post]. www.msci.com/www/blog-posts/frozen-carbon-credit-market-may/ 05232727859.

33. Watanabe, K., Saunders, S., Nishikawa, L., and Turner, G. (2024). *Corporate Emissions Performance and the Use of Carbon Credits*. MSCI Research Insights.

34. Turner et al., "Frozen Carbon Credit Market."

Notes

35. IPCC (2019). *Special Report on the Ocean and Cryosphere in a Changing Climate* (ed. H.-O. Pörtner, D.C. Roberts, V.M. Delmotte, et al.). Cambridge University Press. https://doi.org/10.1017/9781009157964.

36. Huntzinger, D.N., Michalak, A.M., Schwalm, C. et al. (2017). Uncertainty in the response of terrestrial carbon sink to environmental drivers undermines carbon-climate feedback predictions. *Scientific Reports* 7 (4765): www.nature.com/articles/s41598-017-03818-2.

37. It is true: You can in theory drive your car as fast as you want down these highways. And some cars are really fast. In practice, though, most stretches now have some speed limit, usually 74 miles (120 km) per hour, and in many other parts, traffic density does not allow for top speeds. I have also observed recently that fewer and fewer drivers go at "crazy" speeds, because of high gas prices.

38. "BTG Pactual TIG Reaches $500 Million Milestone for Its Latin American Reforestation Strategy" [press release], TIG, November 14, 2024. https://timberlandinvestmentgroup.com/btg-pactual-tig-reaches-500-million-milestone-for-its-latin-american-reforestation-strategy/.

39. Jennifer Ring, Marieta Stefanova, Raiditya Roebino, and Katherine Stodulka. 2023. "The Mangrove Breakthrough Financial Roadmap," December 12, 2023, Scale in Critical Coastal Ecosystems. The Global Mangrove Alliance/Systemiq. www.mangrovealliance.org/wp-content/uploads/2023/11/Mangrove_Breakthrough_Financial_Roadmap_Finance_Coastal_Ecosystems_2023.pdf.

40. Andrew Apampa, *How Much Does a Dollar of Concessional Capital Mobilize?* [blog], February 21, 2023. www.convergence.finance/news/4cC8kVJXvOFZDVxGQ6HLNH/view.

41. World Economic Forum and PwC, *Nature Risk Rising: Why the Crisis Engulfing Nature Matters for Business and the Economy*, WEF, January 2020. https://www3.weforum.org/docs/WEF_New_Nature_Economy_Report_2020.pdf.

42. Gneezy, U. and Rustichini, A. (2000). A fine is a price. *Journal of Legal Studies* 29 (1): 1–17. www.jstor.org/stable/10.1086/468061.

Chapter 5: The World Plus 10 Percent

1. Nemonte Nemquino, "This Is My Message to the Western World—Your Civilisation Is Killing Life on Earth," *The Guardian*, October 12, 2020. www.theguardian.com/commentisfree/2020/oct/12/western-worldyour-civilisation-killing-life-on-earth-indigenous-amazon-planet.

2. World Health Organization, *A Global Health Strategy for 2025–2028: Advancing Equity and Resilience in a Turbulent World*, 14th General Programme of Work. World Health Organization, May 19, 2025. 9789240101012-eng.pdf.

3. Li, S., Keenan, J.I., Shaw, I.C., and Frizelle, F.A. (2023). Could microplastics be a driver for early onset colorectal cancer? *Cancers (Basel)* 15 (13): https://doi.org/10.3390/cancers15133323.

4. Roslan, N.S., Lee, Y.Y., Ibrahim, Y.S. et al. (2024). Detection of microplastics in human tissues and organs: a scoping review. *Journal of Global Health* 14: 04179. https://doi.org/10.7189/jogh.14.04179.

5. Nihart, A.J., Garcia, M.A., El Hayek, E. et al. (2025). Bioaccumulation of microplastics in decedent human brains. *Nature Medicine* 31: 1114–1119. https://doi.org/10.1038/s41591-024-03453-1.

6. National Centers for Environmental Information, "U.S. Billion-Dollar Weather and Climate Disasters," 2024. www.ncei.noaa.gov/access/billions/.

7. Kimmerer, R.W. (2013). *Braiding Sweetgrass. Indigenous Wisdom, Scientific Knowledge, and the Teachings of Plants*. Milkweed Editions.

8. Meadows, D.H., Meadows, D.L., Randers, J., and Behrens III, W.W. (1972). *The Limits to Growth: A Report for the Club of Rome's Project on the Predicament of Mankind*. Universe Books.

9. The fable is usually attributed to Aesop, but its origins are not fully confirmed.

10. Seuss (1971). *The Lorax*. Random House.

11. Kessler, M. and Schmidt-Lebuhn, A.N. (2006). Taxonomical and distributional notes on *Polylepis* (Rosaceae). *Organisms Diversity & Evolution* 6 (1): 67–70. https://doi.org/10.1016/j.ode.2005.04.001.

12. Information from Global Forest Generation Annual Reports, 2020–2024. Available at: www.globalforestgeneration.org/.

13. Frogley, M.R., Chepstow-Lusty, A., Thiele, G., and Chutas, C.A. (2025). Trees, terraces and Llamas: resilient watershed management and sustainable agriculture the Inca way. *Ambio* 54: 793–807. https://doi.org/10.1007/s13280-024-02121-5.
14. Convention on Biological Diversity. www.cbd.int/convention. Retrieved May 31, 2025. The quotation can be found at: https://www.cbd.int/convention/articles/default.shtml?a=cbd-01.
15. Satoyama Initiative, *Case Studies*, n.d. https://satoyamainitiative.org/case_study/#start. Retrieved March 9, 2025.
16. "Coppicing is the traditional method in woodland management of cutting down a tree to a stump, which in many species encourages new shoots to grow from the stump or roots, thus ultimately regrowing the tree. A forest or grove that has been subject to coppicing is called a copse or coppice, in which young tree stems are repeatedly cut down to near ground level." Wikipedia contributors, "Coppicing," *Wikipedia, The Free Encyclopedia*, n.d. https://en.wikipedia.org/w/index.php?title=Coppicing&oldid=1294553244. Retrieved March 22, 2025.
17. European Environment Agency, "Natura 2000 Sites Designated Under the EU Habitats and Birds Directives," December 19, 2024. www.eea.europa.eu/en/analysis/indicators/natura-2000-sites-designated-under?activeAccordion=.
18. European Commission, "Natura 2000 and Forests 'Challenges and Opportunities'—Interpretation Guide." Office for Official Publications of the European Communities, 2003. www.lu.lv/materiali/biblioteka/es/pilnieteksti/vide/Natura%202000%20and%20forests.%20Challenges%20and%20opportunities.pdf.
19. Stevens, C., Winterbottom, R., Springer, J., and Reytar, K. (2014). *Securing Rights, Combating Climate Change: How Strengthening Community Forest Rights Mitigates Climate Change*. World Resources Institute. https://files.wri.org/d8/s3fs-public/securingrights-full-report-english.pdf.
20. Schuster, R., Germain, R.R., Bennett, J.R. et al. (2019). Vertebrate biodiversity on indigenous-managed lands in Australia, Brazil, and Canada equals that in protected areas. *Environmental Science & Policy* 101: 1–6. /doi.org/10.1016/j.envsci.2019.07.002.

21. Rights and Resources Initiative, "Forest and Land Tenure," 2025. https://rightsandresources.org/tenure-tracking/forest-and-land-tenure/.

22. Sze, J.S., Roman Carrasco, L., Childs, D., and Edwards, D.P. (2022). Reduced deforestation and degradation in indigenous lands pan-tropically. *Nature Sustainability* 5: 123–130. https://doi.org/10.1038/s41893-021-00815-2.

23. Graziano Ceddia, M., Gunter, U., and Corriveau-Bourque, A. (2015). Land tenure and agricultural expansion in latin America: the role of indigenous peoples and local communities' forest rights. *Global Environmental Change* 35: 316–322. https://doi.org/10.1016/j.gloenvcha.2015.09.010.

24. Information about these communities can be found in various sources. For Papua New Guinea: The Nature Conservancy, *Women Guardians of the Mangroves*, n.d. www.nature.org/en-us/about-us/where-we-work/asia-pacific/papua-new-guinea-solomon-islands/stories-in-papua-new-guinea-solomon-islands/women-guardians-of-the-mangroves/. For the Skolt Sámi people in Finland, see: Simone Weichenrieder, "Leveraging Indigenous Knowledge for Effective Nature-Based Solutions in the Arctic," The Arctic Institute, Arctic Extractivism Series 2024, August 27, 2024. www.thearcticinstitute.org/leveraging-indigenous-knowledge-effective-nature-based-solutions-arctic/. For information about the Andrafainkona community in Madagascar, see: World Wildlife Fund, "Community Management of Rainforests—Northern Highlands, Madagascar," n.d. https://wwf.panda.org/discover/our_focus/governance/working_with_indigenous_peoples_for_a_fairer_greener_future/#:~:text=In%20Madagascar's%20lush%20Northern%20Highlands,forest%20and%20supervise%20all%20logging.

25. Perhaps "they/them/theirs" would be more appropriate. However, as in this book I am also referring to the concept of Pachamama, a central figure in the cosmology and spirituality of the Indigenous peoples of the Andes often translated as "Mother Earth," it seems more fitting to use "she/her/hers."

26. New Zealand Legislation, *Te Awa Tupua (Whanganui River Claims Settlement) Act 2017*. New Zealand Public Act 2017, No. 7. Date of assent: March 20, 2017. www.legislation.govt.nz/act/public/2017/0007/latest/whole.html.

27. Viktoria Kahui, "Granting Legal "Personhood" to Nature Is a Growing Movement—Can It Stem Biodiversity Loss?," *The Conversation*, April 25, 2024. https://theconversation.com/granting-legal-personhood-to-nature-is-a-growing-movement-can-it-stem-biodiversity-loss-227336.

28. Polanyi, K. (1944). *The Great Transformation*. Farrar & Rinehart.

29. Raworth, K. (2017). *Doughnut Economics: Seven Ways to Think Like a 21st-Century Economist*. Chelsea Green Publishing.

30. Pascual, U., Balvanera, P., Anderson, C.B. et al. (2023). Diverse values of nature for sustainability. *Nature* www.nature.com/articles/s41586-023-06406-9.

Chapter 6: A Trillion Trees

1. Quoted in Global Evergreening Alliance, *Restore Africa: Natural Fighters for Nature* [film], 2024. Melbourne. www.evergreening.org/.

2. The epic of Gilgamesh is recounted in: Langdon, S. (2003). *The Epic of Gilgamesh*. Translated by Andrew R. George. Penguin Books.

3. This saying is often attributed to François-René de Chateaubriand, a French writer, politician, and diplomat who lived from 1768 to 1848.

4. Edwards, P.E.T., Sutton-Grier, A.E., and Coyle, G.E. (2013). Investing in nature: restoring coastal habitat blue infrastructure and green job creation. *Marine Policy* 38: 65–71. https://doi.org/10.1016/j.marpol.2012.05.020. Seymour, Frances, and Jonah Busch. (2016). Why Forests? Why Now? The Science, Economics and Politics of Tropical Forests and Climate Change. Center for Global Development.

5. Bastin, J.F., Finegold, Y., Garcia, C. et al. (2019). The global tree restoration potential. *Science* 5 (6448): 76–79. https://doi.org/10.1126/science.aax0848. Erratum and comments in: *Science* 29, no. 6494 (2020). https://doi.org/10.1126/science.abc8905.

6. Mo, L., Zohner, C.M., Reich, P.B. et al. (2023). Integrated global assessment of the natural forest carbon potential. *Nature* 624: 92–101. https://doi.org/10.1038/s41586-023-06723-z.

7. Crowther, T., Glick, H., Covey, K.R. et al. (2015). Mapping tree density at a global scale. *Nature* 525: 201–205. https://doi.org/10.1038/nature14967.

8. UN Food and Agriculture Organization (2020). *Global Forest Resources Assessment*. Rome: Food and Agriculture Organization of the United Nations. www.fao.org/interactive/forest-resources-assessment/2020/en/.

9. UN General Assembly, "Resolution Adopted by the General Assembly on 1 March 2019," 73rd Session, 73/284. United Nations Decade on Ecosystem Restoration (2021–2030), Paragraph 1. https://docs.un.org/en/A/RES/73/284.

10. FAO, IUCN CEM, and SER (2021). *Principles for Ecosystem Restoration to Guide the United Nations Decade 2021–2030*. Rome: FAO. See the list of principles at: UN Environment Programme, "Panel Unveils 10 Guiding Principles in Campaign to review the Earth," September 8, 2021. www.unep.org/news-and-stories/story/panel-unveils-10-guiding-principles-campaign-revive-earth.

11. The full text of the 10 Principles in the 6 UN languages can be found at this link: https://openknowledge.fao.org/items/8bcc26f1-1a1d-42ce-beb6-2db709d779e6.

12. Benioff, M. (2019). *Trailblazer: The Power of Business as the Greatest Platform for Change*. Crown Currency.

13. A summary of the results can be found at: Salesforce, "100M Trees Funded by 2030: Evaluating Our Progress," June 25, 2024. www.salesforce.com/news/stories/tree-audit-findings-2024/.

14. A lot of guidance for ecosystem restoration across all types of forests and other ecosystems has been published in recent years. For example, some of the best practices are summarized in this blog I wrote for UNEP in 2019:Tim Christophersen, *What Would It Really Take to Plant a Trillion Trees [blog]*, August 23, 2019. www.unep.org/news-and-stories/blogpost/what-would-it-really-take-plant-trillion-trees. Also see: Besseau, P., Graham, S., and Christophersen, T. (ed.) (2018). *Restoring Forests and Landscapes: The Key to a Sustainable Future*. Vienna, Austria: Global Partnership on Forest and Landscape Restoration. www.iufro.org/publications/restoring-forests-and-landscapes-the-key-to-a-sustainable-future.

15. Rinaudo, T. (2022). *The Forest Underground: Hope for a Planet in Crisis*. ISCAST.

16. Werden, L.K., Cole, R.J., Schönhofer, K. et al. (2024). Assessing innovations for upscaling forest landscape restoration. *One Earth* 7 (9): 1515–1528. https://doi.org/10.1016/j.oneear.2024.07.011.

17. Under the Bonn Challenge, for example, a global restoration goal initiated by the International Union for Conservation of Nature and the Government of Germany, 59 governments, private associations, and other entities have pledged to bring 420 million acres (170 million ha) into restoration by 2020 and 860 million acres (350 million ha) by 2030.

18. Will Dean, "'This Machine Kills CO_2 —Inside the 12 July Edition of *Guardian Weekly*," July 10, 2019. www.theguardian.com/news/2019/jul/10/this-machine-kills-co2-inside-the-12-july-edition-of-guardian-weekly.

19. MSCI, *Corporate Emission Performance and the Use of Carbon Credits.* June 1, 2023. www.msci.com/www/research-report/corporate-emission-performance/04624149658.

20. Mann, M.E. (2021). *The New Climate War: The Fight to Take Back the Planet.* Public Affairs.

21. Damian Carrington, "We Asked 380 Top Climate Scientists What They Felt About the Future. . .," *The Guardian*, May 8, 2024. www.theguardian.com/environment/ng-interactive/2024/may/08/hopeless-and-broken-why-the-worlds-top-climate-scientists-are-in-despair.

22. Al Gore, "Political Will Is a Renewable Resource," Speech at the World Economic Forum Sustainable Development Impact Summit, New York, 2018. www.youtube.com/watch?v=V92-TSwJtXY.

23. Harris, N.L., Gibbs, D.A., Baccini, A. et al. (2021). Global maps of twenty-first century forest carbon fluxes. *Nature Climate Change* 11: 234–240. https://doi.org/10.1038/s41558-020-00976-6.

24. Natural Resources Institute Finland, "Preliminary Greenhouse Gas Inventory Results for 2023: Forest Land Has Turned into an Emission Source Because the Carbon Sink of Trees No Longer Cover Emissions from Forest Soil," January 15, 2025. www.luke.fi/en/news/preliminary-greenhouse-gas-inventory-results-for-2023-forest-land-has-turned-into-an-emission-source-because-the-carbon-sink-of-trees-no-longer-cover-emissions-from-forest-soil.

25. UN Environment Programme, *Spreading like Wildfire: The Rising Threat of Extraordinary Landscape Fires*, February 23, 2022. A UNEP Rapid Response Assessment. Nairobi. www.unep.org/resources/report/spreading-wildfire-rising-threat-extraordinary-landscape-fires.

26. UN Environment Programme, "Number of Wildfires to Rise by 50 Per Cent by 2100 and Governments Are Not Prepared, Experts Warn" [press release], February 23, 2022. www.unep.org/news-and-stories/press-release/number-wildfires-rise-50-cent-2100-and-governments-are-not-prepared.

Chapter 7: World Restoration Flagships

1. Levis, C., Costa, F.R.C., Bongers, F. et al. (2017). Persistent effects of pre-columbian plant domestication on Amazonian forest composition. *Science* 355 (6328): 925–931. https://doi.org/10.1126/science.aal0157.

2. Clement, C.R., Denevan, W.M., Heckenberger, M.J. et al. (2015). The domestication of Amazonia before European conquest. *Proceedings of the Royal Society: Biological Sciences* 282 (1812): https://pubmed.ncbi.nlm.nih.gov/26202998/.

3. de Souza, J.G., Schaan, D.P., Robinson, M. et al. (2018). Pre-columbian earth-builders settled along the entire Southern Rim of the Amazon. *Nature Communications* 9 (1125). https://doi.org/10.1038/s41467-018-03510-7.

4. M. Kat Anderson and Michael J. Moratto, *Native American Land-Use Practices and Ecological Impacts*. University of California, Davis. Sierra Nevada Ecosystem Project: Final Report to Congress, 1996. https://babel.hathitrust.org/cgi/pt?id=uc1.31210010519773&seq=1. See also Baylor University, "Native Americans Modified American Landscape Years Prior to Arrival of Europeans," *ScienceDaily*, March 22, 2011. www.sciencedaily.com/releases/2011/03/110321134617.htm.

5. George D. Gann, Tein McDonald, Bethanie Walder,. . .Kingsley W. Dixon, "*International Principles and Standards for the Practice of Ecological Restoration, 2nd ed.*," *Restoration Ecology*, September 4, 2019. https://doi.org/10.1111/rec.13035.

6. The Pact for the Restoration of the Atlantic Forest. www.pactomataatlantica.org.br/o-movimento/#. Retrieved July 1, 2025.

7. UN Decade on Ecosystem Restoration, 2021–2030. *Trinational Pact for the Atlantic Forest*, n.d. www.decadeonrestoration.org/trinational-atlantic-forest-pact. Retrieved July 1, 2025.

8. UN Decade on Ecosystem Restoration, 2021–2030. *Business Enhancement Group: Flagship Finance Report*. Ursula Wilmott, lead author. UNEP, Nairobi. 2024. https://www.decadeonrestoration.org/publications/flagship-finance-project-report.

9. UN Decade on Ecosystem Restoration, 2021–2030. *New UN World Restoration Flagships*. www.decadeonrestoration.org. Retrieved May 31, 2025.

10. Rhett Ayers Butler, "Amazon Deforestation in Brazil Plunges 31% to Lowest Level in 9 Years," *Mongabay*, November 19, 2024. https://news.mongabay.com/2024/11/amazon-deforestation-in-brazil-plunges-31-to-lowest-level-in-9-years/.

11. Catarina Jakovac, Nathália Nascimento, Silvia C. Gallegos,. . .André Pellicciotti, *Strategies for Implementing and Scaling Up Forest Restoration in the Amazon* [policy brief]. Science Panel for the Amazon, UN Sustainable Development Solutions Network, New York. /www.theamazonwewant.org/spa-reports/ https://doi.org/10.55161/DONQ6751.

12. Ministério do Meio Ambiente e Mudança do Clima, Departamento de Florestas, Secretaria de Biodiversidade, Florestas e Direitos Animais, "Plano Nacional de Recuperação da Vegetação Nativa, 2025–2028 [National plan for the recovery of native vegetation, 2025–2028]. Sumário Executivo. Author. 2024. www.gov.br/mma/pt-br/centrais-de-conteudo/publicacoes/biodiversidade-e-biomas/sumario-executivo-planaveg/.

13. World Bank, "Agriculture, Forestry, and Fishing, Value Added (% of GDP)—Brazil," 2024. https://data.worldbank.org/indicator/NV.AGR.TOTL.ZS?locations=ZJ.

14. International Monetary Fund, *World Economic Outlook (October 2024) Data Mapper*. Washington, DC.

15. Wikipedia contributors, "Farmers' Suicides in India," *Wikipedia, The Free Encyclopedia*, https://en.wikipedia.org/wiki/Farmers%27_suicides_in_India. Retrieved April 6, 2025.

16. The next section is based on a blog I wrote while working for UNEP: Tim Christophersen, *Big Potential Benefits from Restoring Spekboom Thicket Ecosystems in South Africa*, June 17, 2019. www.unep.org/news-and-stories/story/big-potential-benefits-restoring-spekboom-thicket-ecosystems-south-africa.

17. van der Vyver, M.L., Mills, A.J., and Cowling, R.M. (2021). A Biome-wide experiment to assess the effects of propagule size and treatment on the survival of *Portulacaria afra* (spekboom) truncheons planted to restore degraded subtropical thicket of South Africa. *PLoS ONE*. https://doi.org/10.1371/journal.pone.0250256.

18. Global Mangrove Alliance, "The Mangrove Breakthrough: A Call to Action," n.d. www.mangrovealliance.org/news/the-mangrove-breakthrough/. Retrieved May 31, 2025.

Chapter 8: Stubborn Optimists

1. Brahma Kumaris is a worldwide spiritual movement led by women, dedicated to personal transformation and world renewal through Rajyoga Meditation. See www.brahmakumaris.com.

2. Powers, R. (2018). *The Overstory*. Norton.. Kingsolver, B. (2013). *Flight Behavior: A Novel*. Harper Perennial.

3. Ghosh, A. (2016). *The Great Derangement: Climate Change and the Unthinkable*. University of Chicago Press.

4. UN Sustainable Development Platform. https://sdgs.un.org/. Retrieved July 1, 2025.

5. For a more in-depth discussion, see Friends of Pando: www.friendsofpando.org/.

6. Rozenn M. Pineau, Andrea Brunelle, Jesse Morris,. . .Zachariah Gompert, "Mosaic of Somatic Mutations in Earth's Oldest Living Organism, Pando." bioRxiv.10.19.619233; https://doi.org/10.1101/2024.10.19.619233.

7. Forest Service, US Department of Agriculture, "Fishlake Basin Recreation Improvement Plan," n.d. www.fs.usda.gov/r04/fishlake/projects/archive/60002.

8. Mariam Issimdar, "Dormant Seeds from Ice Age Pond Project Germinate," BBC, February 22, 2025. www.bbc.com/news/articles/ckg85931d49o.

9. Paul Romer, *Conditional Optimism* [blog], October 8, 2018. https://paulromer.net/conditional-optimism-technology-and-climate/.

10. Khoury, C.K., Sotelo, S., Amariles, D., and Hawtin, G. (2024). *The Plants that Feed the World–Baseline Data and Metrics to Inform Strategies for the Conservation and Use of Plant Genetic Resources for Food and Agriculture.* Rome: FAO. https://doi.org/10.4060/cc6876en.

11. National Office of Statistics, *Family Spending in the UK: Financial Year Ending March 2017.* Her Majesty's Government, 2017. https://blog.ons.gov.uk/2018/01/18/celebrating-60-years-of-family-spending/.

12. António Guterres, "UN Secretary-General's Remarks to the Food Systems Summit," United Nations, September 23, 2021. www.un.org/en/food-systems-summit/news/un-secretary-generals-remarks-food-systems-summit#:~:text=António%20Guterres%2C%20United%20Nations%20Secretary-General%2023%20September%202021,-%20this%20human%20right%20-%20is%20going%20unfulfilled.

13. Crippa, M., Solazzo, E., Guizzardi, D. et al. (2021). Food systems are responsible for a third of global anthropogenic GHG emissions. *Nature Food* 2: 198–209. https://doi.org/10.1038/s43016-021-00225-9. Tim G. Benton, Carling Bieg, Helen Harwatt, Roshan Pudasaini, and Laura Wellesley, *Food System Impacts on Biodiversity Loss: Three Levers for Food System Transformation in Support of Nature.*, Research Paper, UN Environment Programme and Chatham House, 2021. www.chathamhouse.org/sites/default/files/2021-02/2021-02-03-food-system-biodiversity-loss-benton-et-al_0.pdf.

14. Chris Arsenault, "Only 60 Years of Farming Left if Soil Degradation Continues," Reuters, December 5, 2014. www.reuters.com/article/business/environment/only-60-years-of-farming-left-if-soil-degradation-continues-idUSKCN0JJ1R8/.

15. Reiff, J., Jungkunst, H.F., Mauser, K.M. et al. (2024). Permaculture enhances carbon stocks, soil quality and biodiversity in Central Europe. *Communications Earth & Environment* 5 (1): https://doi.org/10.1038/s43247-024-01405-8.

Notes

16. Minasny, B., Malone, B., Mcbratney, A. et al. (2017). Soil carbon 4 per Mille. *Goderma* 292: 59–86. https://doi.org/10.1016/j.geoderma.2017.01.002.

17. *Roots So Deep* is a four-part documentary series by Carbon Nation that explores the world of adaptive cattle farmers and their conventional farming neighbors. See: https://rootssodeep.org/.

18. These figures are cited from various peer-reviewed and published articles. For a full insight into the AMP research and a list of published research, please visit: https://rootssodeep.org/amp-research.

19. Bickel, S. and Or, D. (2020). Soil bacterial diversity mediated by microscale aqueous-phase processes across biomes. *Nature Communications* 11 (116): https://doi.org/10.1038/s41467-019-13966-w.

20. Tree, I. (2024). *Wilding: How to Bring Wildlife Back*. Macmillan Children's Books.

Acknowledgments

I would like to thank my high school friend Tilmann Lahme, author of *The Mann Family*, for giving me the initial idea for writing a book. Many thanks to my literary agent, Jill Marsal, and to Brian Neill and the team at Wiley, for helping me navigate the publication process. My gratitude to Marc Benioff, Suzanne DiBianca, Sunya Norman, Edward Felsenthal, Dan Farber, Max Scher, Naomi Morenzoni, and the entire team at Salesforce for encouraging me to write this book next to my day job. I also benefited from my former colleagues at UNEP who generously shared their experience with the UN Decade on Ecosystem Restoration: thank you, Susan Gardner, Natalia Alekseva, Ivo Mulder, Gabriel Labbate, Jessica Smith, Ann-Kathrin Neureuther, and the Communications team. I would also like to thank the many women, men, youth, and children whose restoration stories and actions have inspired me. There are too many to mention, here are some of them: Achim Steiner, Akanksha Khatri, Andrew Terry, Benki Piyãko, Christabel Reed and the team at Earthed, Constantino "Tino" Aucca Chutas, Dennis Garrity, Diego Saez Gil, Divya Shemanti, Doug McCauley, Eduardo Mansur, Eliane Ubajero, Elizabeth Wahuti, Felix Finkbeiner, Florent Kaiser, H. David Cooper, Hindou Ibrahim, Horst Freiberg, Ibrahim Thiaw, Jennifer Corpus, Jennifer Morris, Jennifer Tauli (Jing) Corpuz, John Lothspeich, Karen Fawcett, Kevin J. Patel, Lewis Pugh, M. Sanjayan, Marie-Claire Graf, Martin Stuchtey, Mette L. Wilkie, Nicole Schwab, Pavan Sukhdev, Rhett Butler, Sadhguru, Satya S. Tripathi, Shyla Ragav, Siddarth Shrikanth, Thomas Crowther, Tony

Rinaudo, Umazi Mvurya, Veena Balakrishnan, Vicky Tauli-Corpuz, Victoria T. Corpus, Vijay Kumar and the team at CBNF, Wanjira Mathaai, Xiye Bastida, and Yugratna Shrivastava.

Huge appreciation and respect to Salina Abrahams for having coined the term and hashtag #GenerationRestoration at a meeting in New York in 2019 in preparation for the UN Decade, and for her review of the chapters, as well as to Florent Kaiser, Justin Adams, Sam Barratt, and Anthony Mills for early reviews, ideas, and encouragement. And I would like to express my love and sincere appreciation to my family for supporting me during many weekends and evenings over the past year to write this book.

I have not let AI write this book for me. However, I have utilized Notebook LLM and Perplexity to summarize research papers, identify additional sources, and aid in structuring my thoughts.

The views and opinions expressed in this publication are those of the author and do not necessarily reflect the official policy or position of Salesforce or any of my previous employers. Any content provided is not intended to malign any religion, ethnic group, club, organization, company, individual, or anyone or anything.

About the Author

Tim Christophersen has been supporting global progress for a stable climate and abundant biodiversity—the life support system of Planet Earth—for over 25 years. His career spans the European Commission, the International Union for the Conservation of Nature and 15 years as a diplomat with the United Nations Environment Programme. From the first attempt to unite the world behind a global climate pact in Copenhagen in 2009, to the groundbreaking Global Biodiversity Framework in 2022 and leading the UN Decade on Ecosystem Restoration 2021–2030, Tim has been at the cutting edge of collective action to save humanity from, in the words of UN Secretary-General António Guterres, our "suicidal war with nature." With global agreements now in place to guide sustainable development, he joined the private sector in May 2022 to accelerate the sustainability revolution by inspiring the world's largest corporations to take more and faster action for climate and nature. As Vice President of Climate Action in Salesforce, a world market leader in artificial intelligence and customer relationship management, Tim works with thousands of companies and public sector institutions on the sustainability transformation. Tim is also on a personal journey spanning three continents over the past two decades, exploring ways to restore our planetary abundance and beauty. He lives with his family on their regenerative farm in Denmark.

Index

A

Acción Andina, 124–135, 195

ActNow, 233

Adaptive multi-paddock (AMP) grazing, 222–223

Africa, 1–2, 26, 75, 162. *See also specific countries*

African elephant, 1–2

AfriCarbon, 204

Agricultural Revolution, 26, 160

Agriculture:

  commodities, 24–25, 43, 66, 97–98, 188

  ecological literacy on, 44

  in extractive model, 217–219

  multistory systems, 189–190

  native bee conservation, 96–98

  reforestation of land used for, 53

  regenerative, 25, 85–86, 112, 189, 198–200, 216–217, 220, 223

  smallholder farms, 46, 198–200, 206, 216–217, 225–227

  soil and, 19–21, 219–224

  water and, 47, 197–198

Agroforestry, 30, 53, 112, 131, 138, 188–189, 191, 192, 226

Almond trees, 96–97

Alnus acuminata, 131

Amazon Forest, 15, 16, 35, 50–52, 54, 59, 139–143, 196–198

Amboseli National Park, 1–2

American bison, 17–18

American Forests, 48, 163

AMP (adaptive multi-paddock) grazing, 222–223

Andes Mountains, 124–135, 173–174

Andhra Pradesh, India, 198–200

Andrafainkona, 142

Antarctica, 16, 35

Anthropocene era, 33
Apollo space program, 3
Arbor Day Foundation, 163
Arc of Restoration, 196–198
Argentina, 52, 124, 184, 187
Aridagawa, Japan, 136
Aristotle, 39
Artificial intelligence, 31, 49, 95,
  165, 171, 189, 200,
  223–224, 229–230
Asociacion Ecosistemas
  Andinos (ECOAN), 127
Assisted natural regeneration,
  172–174, 206
Atlantic Forest of Brazil,
  52, 111, 184–193
Atlantic Ocean, 7–8, 16
Atlantic sturgeon, 10
Aucca, Constantino "Tino"
  Chutas, 125–129, 131
Australia, 2, 17, 86, 111–112,
  141, 176, 180

B
Bald eagles, 26
Bats, 226
Bavianskloof Valley, South
  Africa, 200–205
Bayer, 223
Bees, conserving native, 96–98
Benevolent intelligence,
  230–232

Benioff, Marc, 157, 159, 161,
  165, 166
Bernáldez, Andrés, 4
Betterment, 235
BGT Timberland Investment
  Group, 113
The Biggest Little Farm (film),
  236
Billion Oysters Project, 6–7
Billion Trees Campaign, 158, 159
Biodiversity:
  in extractive model, 86,
    93–94
  in forests, 137–138, 154, 167
  increasing, 6, 74–75, 99, 175,
    177, 227
  loss of, 4–14, 25–26, 44, 65,
    219
  and natural abundance, 3
  soil, 32, 219–225
Biodiversity credits, 109, 114, 176
"Biodiversity Jenga" (Wong),
  59–60
Biomas, 191
Biophilia hypothesis, 43
Biosphere, 33–34, 53, 64, 83, 154
Biotic pump model, 51–52,
  59, 197
Blue mussels, 71
Blue whales, 21–22
Bolivia, 124, 196
Borneo, 58–59, 91, 100

Brahma Kumaris tradition, 209, 214–215
Brazil, 51, 52, 86, 96, 111, 113–115, 141, 174, 184–193, 196–198
Building with Nature, 206
Byck, Peter, 45, 222–223

C
California, 20, 35, 96–98, 180, 225
Canada, 16, 86, 141, 166
Carbon:
  decarbonization, 105, 107, 176
  market for carbon credits, 105–109, 112, 114, 176, 192, 197, 204–205
  sequestration, 22–24, 51, 53, 55, 99, 104–105, 107–109, 157, 179, 212, 220–223
Carbon cycle, 35, 54, 154–155
Carson, Rachel, 77
Cattle, *see* Livestock
Chile, 124
Chimpanzees, xiii–xiv, 77–78
China, 9, 26, 52, 77, 162, 166
Clean Air Act, 94
Climate change, 24, 30, 33–37, 42–43, 53–56, 75, 100, 103–109, 119–120, 175–181
Climate Forest Fund, 85

Cocoa, 24
Collective action, 210, 214–215
Colombia, 124, 196
Common Ground (film), 236
Concessional capital, 115
Conditional optimism, 213, 214
Congo, Democratic Republic of the, 161, 166
Conservation International, 163, 206
Conservationists, 70–77, 138–139, 142, 155–156
Convention on Biological Diversity, 74, 135
Convention on International Trade in Endangered Species, 9
Convergence, 122–123
Coppice forests, 137, 172
Coral reefs, 11–13, 17, 26, 35, 111–112, 219
Cosmovision, 121
Costa Rica, 88, 92–93
Courageous Land, 188–191
COVID-19 crisis, 215–216
Crowding out effect, 118
Crowther, Thomas, 52
Cusco, Peru, 128–129, 133

D
Darwin, Charles, 32
Dasgupta, Partha, 85, 93–94

Da Silva, Luiz Inácio Lula,
    54–55, 196, 198
Death, 144
Denmark, 75, 85, 215–216
Drought, 1–2, 19, 24, 29, 51, 54,
    169, 203, 225
Drought Bird, 122
Ducks, migration of, 10
Dust Bowl disaster, 18–21
Dutch Commonland Foundation,
    103

E
Earth:
    ecological studies of, 31–33
    frameworks for global
        change, 227–228
    impact of climate change on,
        33–37
    planetary-scale restoration,
        32, 53–54, 86, 134, 135,
        183, 206–207. See also
        United Nations, World
        Restoration Flagships
    self-healing powers of, 3–4, 33
Earthshot Prize, 2, 126, 127
ECOAN (Asociacion Ecosistemas
    Andinos), 127
Ecological literacy, 20–21, 40,
    42–56, 101
Ecology, 29–56
    in ancient Greece, 37–41

defined, 31, 86
Gaia hypothesis, 32–33
impact of climate change,
    33–37
interconnectedness in nature,
    31–32
UN Decade on Ecosystem
    Restoration, 41–42
Economic value of nature,
    xvi, 85–118
    Dust Bowl disaster, 20–21
    in extractive model, 86–93
    for finance industry, 99–111
    government incentives and
        investment in restoration,
        85–86, 111–115
    motives for investing in
        nature, 115–118
    price of agricultural
        commodities, 24–25
    in regenerative model, 93–98
Ecopreneurs, 40, 108–109, 196
Ecosia, 233
Ecosystem engineers, 21–24, 71,
    83, 136, 228
Ecosystem restoration. See also
    Forest restoration
    projects; United Nations,
    World Restoration
    Flagships
    on author's farm, 216–217,
        225–227

costs of, 53, 87, 91–93,
111–113
effectiveness of, 27–28
with Indigenous/local
community, xv–xvi,
176–177
investments in, 91–93, 95–98,
111–115
need for, xiv–xv
as our generation's moonshot,
2–4
to reduce wildfire risk, 181
shared responsibility for, 132
Ecosystem Restoration
Communities, 233
Ecosystems, borders between,
58–59
Ecuador, 124, 140, 147, 196
Edge habitats, 136
Enlightenment Era, 61–72, 77,
81–82, 84
Environmental movement, 78,
213–214
Eudaimonia, 39
Europe, 2, 7–10, 16, 26, 27, 220.
See also specific countries
European beavers, 27
European Commission, 137, 139
European eel, 7–9
European Emissions Trading
Scheme, 106
European honeybees, 97

European Union, 85, 94, 98,
137–139, 195, 216–217
Evapotranspiration, 50
Externalities, 94–95
Extinction, 11, 16–18, 26, 65, 80
Extractive model of economy,
66, 82, 86–93, 99,
121–122, 142, 217–219

F
Fantastic Fungi (film), 236
Fernandez, Peter, 196–197
Ferwerda, Willem, 103
Finance industry, 99–109
Finkbeiner, Felix, 158, 159
Finland, 142, 179
Fire Ready Formula, 180–181
Fishing, 8–14, 16–17, 22–25,
71–72, 205, 206
Fishlake Basin, Utah, 212
Food, 29, 30, 44–46, 90–91,
120–122, 184–186.
See also Agriculture;
Fishing
Forced labor, 146–147
Forests, 26, 153–165
in ancient Greece, 37–39
biotic pump model, 51–52
climate mitigation by, 105
drinking water from, 46–47
ecosystem services from,
154–155

Forests (*continued*)
  Indigenous stewardship over,
    141–142
  in Natura 2000 network,
    137–139
  nutrient exchanges in, 31–32
  reasons to save, 155–156
  relationship with, 153–154,
    156–162
  rights-based approach to
    conserving, 162–165
  water cycle in, 50–51
Forest restoration projects,
    157–181
  Acción Andina, 124–135
  in ancient Greece, 37–41
  in Brazil, 113–115
  and climate change, 178–181
  in Costa Rica, 92–93
  criticism of, 175–178
  ecological literacy on, 50–55
  holistic approach to, 169–175
  planting trees in, 172–178
  pragmatic approach to,
    164–165
  rights-based approach to,
    162–165
  Salesforce in, 165–171
  Trillion Trees Initiative,
    156–162, 165–168
Fossil fuels, 36, 52, 91–93, 217,
    221, 223, 225

4Returns Framework, 103
Francis, Pope, 57, 68, 79–80, 210
French Guiana, 196
Fuller, Buckminster, 3, 42

G

Gaia hypothesis, 32–33
Germany, 10–11, 23, 71–72,
    110–111
Ghosh, Amitav, 210
Global Biodiversity Framework,
    74, 76, 109
Global Biodiversity Fund, 181
Global Evergreening
    Alliance, 30
Global Forest Generation, 126,
    129, 130
Global Mangrove Breakthrough,
    114, 165, 205–206
Global Partnership on
    Forest Landscape
    Restoration, 96
Global warming, 33–37, 48,
    52–53, 104, 175–176, 178,
    214–215
Gluesenkamp, Dan, 97
Gneezy, Uri, 117–118
Goats, *see* Livestock
Goodall, Jane, 77–78
Good On You, 233
Google, 107
Gore, Al, 179

Government:
fossil fuel subsidies, 91–93
investment in restoration, 102,
109–115, 140
land protection by, 71–75,
137–139, 141–142,
164, 184
Grant capital, 113, 130
Grazing, 222–223. *See also*
Livestock
Great Barrier Reef, 17, 111–112
Great Green Wall, 206
Great Plains, 17–21
Great Plains Shelterbelt project,
21
Greece, ancient, 37–41
Green Belt Movement, 157
Green Belt of Metropolitan
Watersheds, 96
Greenhouse gas emissions,
xiv, 23, 36, 44, 108, 110,
162, 192, 219
Greenland, ice sheet collapse, 16
Greenwashing, 167, 177, 223
GRID-Arendal, 180–181
Guatemala, 141
Guterres, António, 77, 80
Guyana, 196

H
Harvest restrictions, 120–122
Hawkmoths, 225

Herring migration, 10
Hichilema, Hakainde, 29
Homo sapiens, 62, 66
Honorable Harvest, 120–121
Hudson-Raritan Estuary, 5–7
Hughes, Terry, 17
Human Microbiome Project, 63
Human rights, 146–147
Hunting restrictions, 120–122

I
iNaturalist app, 49
Inca Empire, 127, 128, 131–132
India, 166, 198–200
Indigenous and local
communities, 119–151.
See also specific projects
conservationists' views of,
70–71
convergence for, 122–123
ecological literacy of, 43
ecosystem restoration with,
xv–xvi, 176–177
relationship with nature for,
81–82, 143–151
stewardship by, 139–143
wars against, 17–18
Indonesia, 90–91, 205–206
Indoor Generation, 98
Industrial Revolution, 5, 145
Ingham, Elaine, 221–222
Insect diversity, 225

Institute and Faculty of
Actuaries, 89
International Forestry Students
Association, 156
International Monetary Fund,
67, 91
International Union for
Conservation of
Nature, 7, 75

J
Jane Goodall Institute, xv, 171
Japan, 9, 86, 98, 135–136
Java, 206
JouleBug, 234

K
Kaiser, Florent, 126, 127, 129,
132, 134–135, 156
Kariba Dam, 29
Karura Forest, Nairobi, 157–158,
164–165
Kauders, Phil, 191
Kenya, 47, 76
Key West, Fla., 11–13
Kimmerer, Robin Wall,
29, 120
Kiss the Ground (film), 236
Knepp Estate (Sussex, England),
224
Kruger, Pieter, 201–202, 205
Kurlansky, Mark, 5

L
Landscape approach, 143–148,
169–175, 177–178
Laos, 174
LeafSnap app, 49
Lidong Mo, 52–53
Lima, Peru, 125
Livestock, 44–45, 137–139,
173–175, 186–187,
190–191, 201, 203–205,
212, 216, 222–223, 227
Living Indus project, 206
Local communities, see
Indigenous and local
communities
Local produce, 44–46
Long Peace, 229
Los Angeles, Calif., 2025
wildfire, 180
Lovelock, James, 32

M
Maathai, Wangarĩ, 157–158
McClenachan, Loren, 11–13
Mackinnon, J. B., 26
Madagascar, 142, 169, 170
Mangroves, 112, 114, 142,
169–170, 205–206
Mann, Michael E., 178
Mao Zedong, 77
Marais, Christo, 204
Mass extinction events, 11, 65, 80

Matico plant, 140
Meat consumption, 44–45
Medicinal plants, 140, 167, 185
Mediterranean Forest
    Landscapes, 206
Merlin app, 49
Mexico, 141
Microbiota, 63–64
Microplastics, 119
Microsoft, 107
Mills, Anthony, 203
Mind-body problem, 67–68
Mining, 9, 130
Mombak, 191, 197
"Mother Nature," 143–148
Mushroom ID app, 49

N
Naidu, Nara Chandrababu, 199
Nairobi, 157–158, 164–165
National Peatland Strategy, 111
Natura 2000 network, 137–139,
    172
Natural abundance, 3, 6, 14,
    25–27, 93, 214
Natural capital, 69–70, 87–90, 93,
    99, 100
Nature. *See also* Economic value
    of nature; Relationship
    with nature
    as asset class, 99–103,
        112–113, 168

business case for, 115–116
climate mitigating solutions
    from, 54–56, 103–104
as critical infrastructure,
    34–35, 109–111, 183
defining, 64–65, 81, 84
deification of, 70–73
direct dependencies on,
    48–49
flywheel of life in, 6–7, 224
framework for closeness
    with, 149–150
interconnectedness of,
    31–32, 58–60, 121
resilience of, 14–15, 24–28,
    34, 59–60, 123, 167–168,
    211–213
spending time in, 150–151
spiritual aspects of, 230–232
strength of, 144–145
tipping points in, 14–24, 35
Nature Conservancy,
    163, 206
Nature investment funds,
    101–103
The Nature of Farming (film),
    236
Nenquimo, Nemonte, 82, 119
New York Harbor, 4–7, 15
New Zealand, 147
Niger, 173
99 (startup), 197

Nitrates, 47
Nitrogen, 219, 221–223
Nobre, Carlos, 196
North Atlantic cod, 16–17

O
Ocean (film), 236
Oceans, 17, 26, 108, 219
Ojibwe people, 120
Oliver, Tom, 81
Optimism, 209–228
    about relationship with
        nature, 215–227
    envisioning future with,
        210–215
    for making change, 27–28,
        227–228
    from religion and literature,
        209–210
Oysters, 4–7, 15

P
Pact for the Restoration of the
    Atlantic Forest (Pacto
    Mata Atlântica), 187,
    188, 192–193, 195
Pakistan, 206
Pando (aspen tree system),
    211, 212
Papua New Guinea, 142
Paraguay, 184, 187
Paraná pine, 190

Paris Agreement on Climate
    Change, 26, 104, 110,
        165, 181, 214–215, 221
Pascual, Unai, 149
Patagonia Action Works, 234
Peatlands, 90–91, 110–111
Perlin, John, 37
Permaculture, 217, 220, 227
Peru, 124, 128–129, 133,
        134, 196
Pesticides, 77, 92, 97, 199, 217,
        219, 221, 223, 226
Peterson, Heidi C., 23
Philosophy, 38–39, 67–70, 82
Phytoplankton, 22, 23
Pine trees, 144, 190
Pingos, 212–213
Planetary Boundaries, 31
Plant-for-the-Planet, 158, 159,
        163, 234
Plato, 38, 39
Polanyi, Karl, 147–148
Pollution, 94–95, 98, 119
Polylepis trees, 125, 128–129,
        133, 173–174
Predators, reintroduction
    of, 212
Princess Mononoke (film), 236
Private sector:
    climate solutions from,
        104–109, 165–166
    nature finance from, 103–109

278

Project Drawdown, 234
Public-private partnerships, 85, 102, 188

R

Raworth, Kate, 148
Reforestation, *see* Forest restoration projects
Refugio José Rivas, 124
ReGeneration, 234
Regenerative agriculture, 25, 44, 85–86, 112, 189, 198–200, 216–217, 220, 223
Regenerative model of economy, 87, 93–98
ReGreen, 191
Relationship with nature, 57–84
    biological science on, 62–64
    commodity approach in, 2n*, 61, 66, 69, 78, 79, 82, 144, 146, 148
    conservationists' views of, 70–77
    crisis in, 30–31
    culture as external to nature, 64–65
    ecological literacy on, 55–56, 58
    in "Enlightenment" Era, 61–72
    forests and, 153–154, 156–162
    holistic view of, 57–60

humans as part of nature, 57, 60–64, 71–72, 75, 78–84
humans as separate from nature, 64–79
for Indigenous and local communities, 81–82, 143–151
language describing, 64–65, 80–81
peaceful, 77–78, 80–81, 145
reciprocal, 40, 70, 77, 79, 82, 122, 127, 128, 135–136, 143, 146–147, 150
religion and philosophy on, 65–70, 79–80, 82
resetting/restoring, 77–82, 156–162, 175, 178, 215–227
societal breakdown over, 29–31
Religion, 61, 66–67, 79–80, 82, 83, 121, 209–210
Renewable energy, 29, 47, 52, 129, 132, 178–179
Responsibility, 117–118, 132
Restor, 171, 177, 234
Restoration economy, 191–192, 197, 200–205
Rights-based approach, 74, 146–147, 162–165, 192
Right whales, 4
Rinaudo, Tony, 173

Rio de Janeiro, Brazil, 186
Rome, ancient, 146
Romer, Paul, 213
Roots and Shoots, 234
Roots So Deep (film), 45, 222–223, 237
Rustichini, Aldo, 117–118

S

Safina, Carl, 69
Sahel region, 172–173, 206
Salesforce, 104–105, 107, 114, 156, 157, 159, 165–171, 177, 205–206
Salmon, 10
Samoa, 206
Sangre de drago plant, 140
São Paulo, Brazil, 96, 186, 188–191
Sargasso Sea, 7–8
Satoyama Initiative, 135–136
Schmidt, Johannes, 7
Seafood Watch, 234
Sea level rise, 145
Sea turtles, 4
SEEA (System of Environmental-Economic Accounting), 88
Sheep, see Livestock
Shifting baseline syndrome, 11–14
Silva, Gabriel, 197

Simard, Suzanne, 31–32
Sinai desert, 193–194
Skolt people, 142
Smits, David, 18
Society for Ecological Restoration, 186
Socio Bosque, 140
Soil, 19–21, 32, 44, 90–91, 108, 131, 186–187, 189, 202, 219–225
Soil Food Web, 220–224, 235
SOS Mata Atlântica initiative, 187
South Africa, 200–205
South America, 26, 50, 197–198. See also Andes Mountains; specific countries
Spekboom trees, 202–205
Sperm whales, 23
Sri Lanka, 111
Steiner, Achim, 159
Stewardship, 67, 83, 123, 132, 139–143, 161, 185–186
Stop Talking, Start Planting!, 158, 159
Sukhdev, Pavan, 89
Suriname, 196
Sylt, Germany, 71–72
Symbiosis Coalition, 107
System of Environmental-Economic Accounting (SEEA), 88

T

Task Force on Nature-related
Financial Disclosure,
101
Thicket ecosystem, 201–205
30 by 30 plan, 73–76
To Which We Belong (film), 237
Tree, Isabella, 224
Trees, in urban areas, 48–49
Trillion Trees Initiative, 126,
156–162, 165–168, 174,
177, 181
Tropical Forests Forever Facility,
54–55
Turkey, 176

U

UBS, 114
Uganda, 171
Ultra-high-net-worth
conservationists, 76–77
UNESCO World Heritage Sites,
17, 200
United Kingdom, 48, 86, 89,
98, 212–213, 218
United Nations:
Carbon Offset Platform, 235
Decade on Ecosystem
Restoration, xvi, 41–42,
44, 55–56, 80–81, 156,
159, 162, 165, 188,
194–195, 235

Declaration on the Rights of
Indigenous People,
162–163
Environment Programme,
xvi, 55, 92, 135, 156,
158, 159, 165, 180–181,
194, 215
Food and Agriculture
Organization, 194,
219–220
Food Systems Summit,
218–219
Framework Convention
on Climate Change,
156, 163
Intergovernmental Panel on
Climate Change, 31
Intergovernmental
Science-Policy Platform
on Biodiversity and
Ecosystem Services,
26, 80, 91, 149
Programme to Reduce
Deforestation and Forest
Degradation, 140
Statistics Division, 87–88
Sustainable Development
Goals, 171, 210–211
World Restoration Flagships,
183–207
United States, 17–18, 26, 48, 86,
98, 112, 162, 166

Universal Declaration of
    Human Rights, 162
Universal human rights, 146–147,
    162

V
Vanuatu, 206
Venezuela, 124, 196
Von Wong, Benjamin, 59–60

W
Wasai plant, 140
Water, drinking, 46–47, 94–96,
    124–125, 168, 188, 205
Water cycle, 36, 50–51, 154–155
Water utilities, restoration
    projects by, 95–96
Wealthsimple, 235
We Are Nature campaign, 81
We Don't Have Time, 235
WeForest, 163
Western worldviews, 82,
    119–123, 143. *See also*
    Enlightenment Era

Whales, 4, 21–24
Whanganui River, 147
Wildfires, 35, 180–181
William, Prince of Wales,
    2, 126, 127
Wilson, E. O., 43
Working for Water programme,
    202
World Bank, 67, 198
World Economic Forum,
    116, 126, 156, 157,
    159, 161, 163, 165,
    181
Worldwide Fund for
    Nature, 138
World Wide Opportunities on
    Organic Farms
    (WWOOF), 235

Z
Zambezi River, 29–30
Zambia, 29–30, 33
Zoological Society of
    London, 27